The Structure of the Universe

JAYANT NARLIKAR

The Structure of the Universe

OXFORD UNIVERSITY PRESS

Oxford University Press, Walton Street, Oxford, OX2 6DP

LONDON   OXFORD   NEW YORK
GLASGOW   TORONTO   MELBOURNE   WELLINGTON
CAPE TOWN   IBADAN   NAIROBI   DAR ES SALAAM   LUSAKA   ADDIS ABABA
KUALA LUMPUR   SINGAPORE   JAKARTA   HONG KONG   TOKYO
DELHI   BOMBAY   CALCUTTA   MADRAS   KARACHI

Paperback ISBN 0 19 289082 4
Hardback ISBN 0 19 217653 6

© Oxford University Press 1977

First published in OPUS 1977
Reprinted with corrections 1978

Set in Great Britain by Gloucester Typesetting Co. Ltd.
and printed by Fletcher & Son Ltd, Norwich

Preface

Describing a technical subject in a non-technical form is no easy job; but, of all the branches of science, astronomy lends itself most easily to popularization. The writer of a popular book on astronomy starts with the advantage that his reader, more often than not, has already developed a curiosity about the subject. Merely by raising his eyes to the star-studded sky at night, even the layman may have the satisfaction of knowing that he has made an astronomical observation.

For the professional astronomer, however, the days of simply looking through a telescope are gone. Even if equipped with a good optical telescope, he will think of using the instrument in conjunction with a photographic plate or an image tube or some other invention of modern science, rather than employ his naked eye. In the last three decades new branches of astronomy have sprung up in which the telescopes used are totally different from the layman's concept of a tube with lenses at the ends. But, however different in shape, and however modern in technology, these instruments are designed with the same aim: to explore the far corners of the Universe.

Remarkable though this growth in astronomical techniques has been, it is matched on the theoretical side by the astronomer's attempts to interpret his observations. Here he uses daring extrapolations of his knowledge of the laws of science gained in terrestrial laboratories. On occasion he has to go beyond mere extrapolation and introduce new concepts which lie outside the scope of laboratory science. In attempting such radical flights of thought the astronomer is guided by the hope not only of understanding what goes on 'out there', but also of learning a little more of the laws of physics that govern these events. Astronomy has made useful contributions to science in the past and is expected to continue doing so in the future.

While trying to understand the mysteries and grandeur of the Universe in terms of laws of science which are striking in their simplicity and beauty, the astronomer may well recall the words of the mathematician and philosopher Leibniz (1646–1716): 'It follows from the supreme perfection of God that in producing the Universe He chose the best possible plan, containing the greatest variety, together with the greatest order; the best arranged situation, place, and time . . .'

To the layman living on this poverty-ridden, pollution-infested Earth, these words may sound Utopian. But even the Earth does enjoy a rather pleasant situation in the Universe. It is surrounded by a protective atmosphere, it is located at just the right distance from the Sun to enjoy tolerable temperatures, and it has the life-sustaining elements in proper abundance. Is it not fortunate that we are located far from the madding crowd of the galactic centre? Imagine what would happen if our Galaxy were the seat of the type of violent activity seen in the not-too-distant radio source, Centaurus A. Finally, we may owe our existence to the fact that ours is not an epoch of high temperature like the one which many astronomers think the Universe had at the time it was born.

What is the Universe like beyond the protective confines of the Earth? In this book I have tried to describe some of the important aspects of the Universe which have been revealed through astronomical observations, and how the astronomer has tried to explain these in terms of what he knows. Needless to say, there are a number of questions which remain unanswered. But the few questions that have been answered inspire confidence that we may be stumbling and groping along the right path.

The main part of the text is of a non-technical nature. Technical terms have been explained in an extensive glossary at the end of the book, for the benefit of those readers with little scientific background. To elaborate certain points in the main text I have adopted the method of 'boxes'. Those marked with a dagger are of a technical nature and require a knowledge of mathematics and physics up to undergraduate level. My purpose in introducing such material is to explain to the reader with a general scientific and mathematical background how some of the far-reaching results in astronomy can be obtained, at least approximately, by using simple mathematical methods; for example, the central temperature in the Sun, the energy stored in an extragalactic radio source, and the age of the big-bang Universe can all be estimated in this way (as shown in the appropriate boxes). A reader without the necessary scientific background may skip these boxes without losing the main thread of the argument.

I am grateful to my father, Professor V. V. Narlikar, for reading the first draft of the book so carefully and for making numerous constructive suggestions. I thank my colleagues Mahendra Vardya, Krishna Appa Rao, Kumar Chitre, and S. Ramadurai, and my wife Mangala, for reading portions of the manuscript and offering critical comments. Professor Devendra Lal and Professor Vainu Bappu very kindly supplied some of the photographs. To them and to the Hale Observatories I am especially indebted for the photographs appearing in this book. Finally, thanks are due to Mr. P. Joseph for typing the manuscript and to the Drawing Office staff of T.I.F.R. for assistance with the diagrams.

This book forms part of the project undertaken by me as a Jawaharlal Nehru Fellow (1973–5). It is a pleasure to acknowledge the assistance I have received from the Jawaharlal Nehru Memorial Fund during this period.

Jayant Narlikar

Tata Institute of Fundamental Research
Bombay, India

Contents

1

The Astronomer's Universe

Once upon a time an astronomer, a physicist, and a mathematician set off on a walking tour in the Scottish highlands. They soon came across a sheep grazing all alone on a farm. Looking at it the astronomer commented 'So, the sheep in the highlands are black.'

'You cannot generalize so sweepingly', admonished the physicist. 'Your sample is too small. Only after a careful analysis of a large number of sheep all over the highlands can you make such a statement. Just now all you can say is that black sheep are found in Scotland.' He turned to the mathematician for his views.

'I am afraid I disagree with you both', remarked that worthy. 'All you can say is that the animal over there appears to be black on the side facing us.'

The Universe, by definition, includes everything. Even with all his limitations, man has looked at it from many different points of view. The Universe presents one aspect to the philosopher, another to the scientist, and still another to the artist. It is beyond the capacity of any one person to describe the Universe in all its manifestations; and my purpose in these chapters is to limit the description to a scientific one. Even here a further qualification is necessary. I shall be concerned with the large-scale structure of the Universe; that is, the Universe as studied by the astronomer using the various observational techniques at his command and the interpretations based on the laws of science *known today*.

It is important to emphasize the last two words. Our present picture of the Universe necessarily depends on our present knowledge, and there is no reason to believe that this knowledge has reached perfection at the time of writing this book. Indeed, if history is any guide, man's view of the Universe has changed steadily with time, and will continue to do so in the future. Primitive man looked at the heavens with awe and wonder. Instinctively he ascribed divine or mystical powers to the Sun and the other objects in the sky. This gradually gave way to a spirit of rational enquiry, based on a more scientific way of thinking. As the scientific approach to heavenly objects progressed, the mystical element began to give way. In the early days progress was hampered by various social and religious dogmas. The 'take-over' by science, which essentially started with Nicolaus Copernicus (1473–1543) and was completed during the lifetime of Isaac Newton (1643–1727), speeded up the growth of astronomy. Nevertheless, time and again, the astronomer has had to revise his earlier assumptions and readjust his view of the Universe. Whether he likes it or not, a pragmatic outlook has been forced upon him.

The reason is not difficult to see. It lies in the vastness of the Universe, in the enormous quantity of matter it contains, and in the fantastically long time scales its various operations involve. It is just as well to remind ourselves of the large numbers required to describe the Universe. To begin with, our Earth forms part of the Solar System, which consists of the Sun, the planets, their satellites, the asteroids, the meteorites, the comets, and a certain amount of debris. The Earth is about 150 million km away from the Sun. The distance of the furthest planet—Pluto—from the Sun is nearly 40 times as much. Even at this stage the distances are so large that the astronomer does not find it useful to employ the earthbound kilometre as a unit. It is easier to specify the distance in terms of the travel time of light. Light travels at nearly 300 000 km s^{-1}; so it covers the Earth–Sun distance in a little over 8 minutes. It would take nearly $5\frac{1}{2}$ hours to go from the Sun to Pluto. The time taken by light to reach other stars lying near the Sun is of the order of a few *years*. So for measuring star distances the 'light-year', that is, the distance covered by light in one year, is an appropriate unit.†

However, even this unit is inadequate to measure the distance to remoter parts of the Universe. Our Sun belongs to a bigger system, the Galaxy, which contains about a 100 000 million ($= 10^{11}$) stars. The Galaxy is a bun-shaped object with a diameter of nearly 100 000 light-years. The Sun itself lies at a distance of about 30 000 light-years from the centre of the Galaxy, thus not occupying any particularly

† The professional astronomer prefers to use another unit called the 'parsec', which we shall encounter in the next chapter. This is approximately equal to 3 light-years.

privileged position. Nor can we take pride in our Galaxy having any special status in the Universe. It is estimated that there are nearly 3000 million ($= 3 \times 10^9$) other galaxies in the observable Universe, that is as far as our best telescopes can probe. And this distance is as enormous as 10 000 million ($= 10^{10}$) light-years!

And what about masses? Adult human beings generally have masses in the range 50–100 kg. The Earth has a mass of nearly 6 million million million million ($= 6 \times 10^{24}$) kg. The Sun is about 300 000 times as massive as the Earth. Again, the kilogram is too small a unit to describe astronomical masses. A more convenient unit is the mass of the Sun, denoted by the symbol $M_\odot$. Our Galaxy has a mass of about $10^{11} M_\odot$, that is, it is about 100 000 million times as massive as the Sun. In Chapter 2 we will encounter other massive objects far more dramatic in behaviour than our Galaxy. The total amount of matter estimated in the observable Universe is at least 300 million million million solar masses ($= 3 \times 10^{20} M_\odot$)!

The time scales are also very long. The Earth takes a year to complete one orbit round the Sun. The stars in the Galaxy may take as long as 200 million (2×10^8) years to complete one orbit round the Galactic centre. The large-scale structure of the observable Universe changes over time scales of the order of 10 000 million ($= 10^{10}$) years.

Apart from this intimidating catalogue of vastness, the limitation of the human brain may also stand in the way of attaining perfect knowledge of the Universe. The biological make-up of the brain is such as to limit its ability to assimilate and interpret information beyond a certain degree of complexity. There is no reason to believe that the nature of the Universe has just about the same degree of complexity as that which can be understood by the best human brain. It is worth quoting the views of two well known astronomers of the present century. Sir Arthur Eddington compared man, in his search for knowledge of the Universe, to a potato bug in a potato in the hold of a ship trying to fathom from the ship's motion the nature of the vast sea. Whatever method the potato bug may adopt in its experiment, man may claim to have a more scientific approach starting with the basic laws of physics. However, in the year 1970, Sir Fred Hoyle made the following remark in this connection: '... I think it is very unlikely that a creature evolving on this planet, the human being, is likely to possess a brain that is fully capable of understanding physics in its totality. I think this is inherently improbable in the first place, but, even if it should be so, it is surely wildly improbable that this situation should just have been reached in the year 1970.'†

The above remarks might convey the impression that the problem of understanding the Universe is so formidable that it is not worth pursuing at all. This is not the case. Indeed a proper scientific approach to the study of the Universe has already yielded results which indicate how exciting the subject is. Science always has new and challenging problems in store for man, who responds to them by his ingenious theories, experiments, and observations. I hope I can convey through this book some of the excitement involved in accepting the challenge of the Universe.

Astronomy and physics

What is the scientific approach to astronomy? A scientist working in a laboratory conducts his investigations in three steps: (1) experiment, (2) observation, and (3) deduction. Of these, the first step is more-or-less absent in astronomy. The astronomer is handicapped in the sense that he cannot manipulate at will the arrangements and behaviour of heavenly bodies. They are too massive and too far away to be susceptible to his influences. For example, he can study the Sun in some detail because it happens to be the nearest star around. But he is not in a position to say that, having studied the Sun, he would now like to investigate a star half as massive as the Sun by, say, removing half the matter from it. He will have to look around for a star of that mass elsewhere in the Galaxy, and such a star may not be easy to find. The astronomer's counterpart, the laboratory physicist, on the other hand, can design his equipment and vary his experimental parameters in any number of ways to enable himself to check a given scientific law to his satisfaction. The astronomer has to be content with remote observations only, and on occasions his deductions are based on rather scanty data.

When astronomical observations have yielded sufficient data, those data are analysed to discover the underlying pattern. This, again, is not an easy job. The raw data often look extremely messy. The detection of a pattern requires experience, careful analysis, and often an eye of faith. But once the pattern has been discovered, it then

† See *Study Week on Nuclei of Galaxies* (ed. O'Connell), p. 757. North Holland, Amsterdam.

becomes the job of the theoretician to explain the pattern in terms of some basic law of nature. A notable example is that of planetary motions.

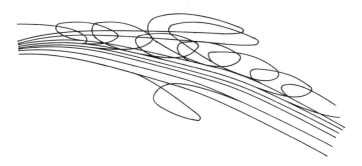

FIG. 1.1. Schematic picture of the simulated planetary motion over several years. The pattern is complicated by the loops and the retracting motions.

Fig. 1.1 shows a picture of simulated planetary motions over a period of several years. Clearly no pattern seems to exist. Yet out of such data, collected over centuries, the Greek astronomers, notably Hipparchus (190–120 B.C.) and Ptolemy (A.D. 85–165), were able to detect 'epicyclic motions' (movements in circles). In Ptolemy's view the Earth was supposed to be at rest while the Sun and the planets moved round it. These motions, in their simplest form, were in the form of circles whose centres moved around on certain other circles (see Fig. 1.2). To represent the motion of a planet fairly accurately sometimes more circles (epicycles) were added. While this pattern was reasonably successful in predicting the future position of a planet in the sky, it was too complicated to reveal any underlying law of nature. Several centuries later Copernicus found another pattern which worked equally well but which was easier to understand. In this the Sun was at rest and the Earth and other planets revolved round it (see Fig. 1.3). Later Johann Kepler (1571–1642), after a detailed analysis of the motion of the planets, found that they moved in elliptical orbits with the Sun at one of the foci. He was also able to give details about how the planets moved along these orbits (see Fig. 1.4). It was by using this pattern that Newton was able to detect

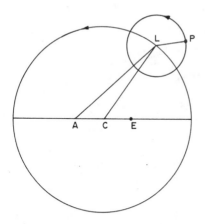

THE CONSTRUCTION OF PTOLEMY

FIG. 1.2. E represents the Earth and P the planet. The line AL turns with the mean angular velocity of the planet round the Sun while the line LP turns with the mean angular velocity of the Earth. The lengths AC (= CE) and CL are specified for each planet, while the length LP is related to the Earth–Sun distance. A somewhat different construction is given for Venus.

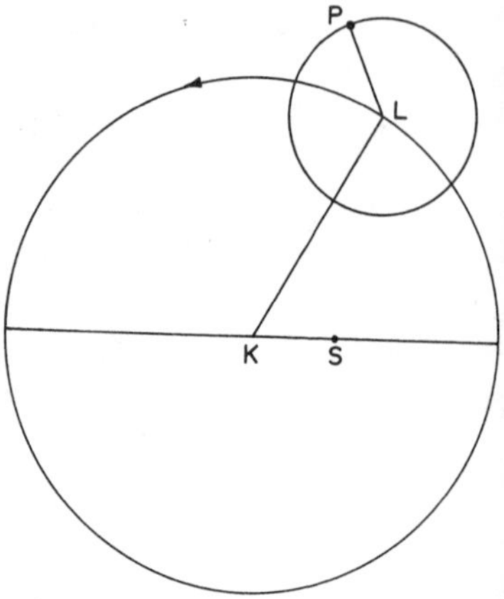

THE CONSTRUCTION OF COPERNICUS

FIG. 1.3. S represents the Sun and P the planet. The line KL turns with the mean angular velocity of the planet round the Sun while LP turns at twice this rate. The lengths KL and KS (= 3LP) are specified for each planet. (Not drawn to scale.)

the underlying natural law: the inverse-square law of gravitation. He was able to demonstrate that this gravitational attraction causes planets to move round the Sun in elliptical orbits. The central role played by the Sun was obvious in the Copernican picture, but had

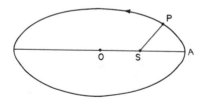

THE KEPLERIAN ORBIT

Fig. 1.4. The line SP joining the Sun to the planet sweeps out equal areas in equal time intervals. OA is half the major axis of the ellipse and the ratio OS/OA is its eccentricity.

not been so obvious in the earlier Greek version; and we see how it had taken man many centuries to reach a true interpretation of the pattern of planetary motion.

What laws of nature does the theoretical astronomer look for in explaining the various observed phenomena? Basically they are the laws of science detected in the laboratory, but with one difference. Astronomical systems are vast. The physical conditions relevant to these systems are often much more extreme than those created in a laboratory experiment. Therefore the laws used by the astronomer have to be extrapolations of laws discovered in the laboratory. For example, the law of gravitation as given by Newton was

$$F = G \frac{m_1 m_2}{r^2}. \tag{1.1}$$

Here F is the force of attraction between two objects of masses m_1 and m_2 situated a distance r apart, and G is the constant of gravitation. This law has been well tested and found satisfactory in the laboratory, and the value of G has been determined. Naturally, the masses m_1, m_2 in the laboratory are small, say measurable in kilograms. However, the astronomer must apply this law to stars and planets, which are much more massive.

As an example of the astronomer's powers of extrapolation consider the legendary apple whose fall is said to have inspired Newton with the law of gravitation. Suppose we set $m_1 = $ mass of the Earth and m_2 = mass of the apple. The distance r in the formula (1.1) is very nearly the radius of the Earth, since it represents the distance of the apple from the centre of the Earth. Then formula (1.1) represents the force of attraction of the Earth on the apple, which causes a downward acceleration of the apple. Denoting the acceleration by the symbol g, we have, from Newton's second law of motion (force = mass × acceleration)

$$F = m_2 g. \tag{1.2}$$

From (1.1) and (1.2) together we arrive at the result

$$m_1 = \frac{gr^2}{G}. \tag{1.3}$$

In this formula g and G are measured by the laboratory physicist and r is known from measurements of the Earth. So from (1.3) the astronomer is able to calculate the mass of the Earth!

Sometimes these extrapolations are clear-cut; sometimes they are not. This has led to the introduction of a certain speculative element into astronomical theory. We have already seen that, from an observational point of view, the astronomer is at a serious disadvantage compared with the laboratory physicist. Because his experimental material is so large and inaccessible his data are not as accurate as the data in a laboratory experiment. For this reason, although astronomy forms part of physics, it has always remained a little apart from the rest of the disciplines in physics.

The relationship between astronomy and physics has fluctuated over the centuries. At the time of Galileo (1564–1642) and Newton, astronomy dominated the scene. There was not much of laboratory physics. With the discovery of electromagnetic phenomena, laboratory physics began to come into its own, and made rapid progress in both the eighteenth and nineteenth centuries. The two revolutions of scientific thought which occurred early in the present century—the special theory of relativity and the quantum theory—came about through laboratory physics. Meanwhile the impact of astronomical discoveries on physical ideas was gradually diminishing, and by the beginning of the present century astronomy had been relegated to the status of a poor relation in the family of sciences. From this neglected state it has begun to emerge again over the last two decades or so. This is largely for two reasons. First, thanks to the recent rapid progress in technology the observational astronomer can now study the Universe in many new ways not previously possible. This is reflected in the emergence of radio-astronomy, X-ray and γ-ray astronomy, infrared astronomy, and so on. Even the optical astronomer has several sophisticated instruments at his disposal which make his data much more interesting and accurate than that of his predecessors. The second reason for the new advances in astronomy lies on the theoretical side. The theoretician has drawn liberally upon the advances made in the fundamental sciences to explain several unusual astronomical phenomena. And so the gap between astronomy and the rest of physics is now closing.

In the next three chapters I wish to highlight some of the successes as well as the difficulties of present-day astronomy. And we shall see how far the extrapolation of laboratory physics has been useful to astronomy. But I close this chapter by posing a different question. To what extent is the large-scale structure of the Universe relevant to the laboratory experiments we perform on the Earth? In other words, is the local behaviour of matter influenced by what goes on in the Universe thousands of millions light-years away? This idea may sound fantastic, but as the following example shows, it has to be taken seriously.

The Olbers paradox

Why is the sky dark at night? In 1826, the Viennese physician and astronomer Heinrich Olbers began to consider this somewhat obvious question, and as a result formulated what is now called the 'Olbers paradox'. On the face of it the question seems ridiculously simple to answer. At night we are facing away from the Sun, and so we look at the part of the sky which is not lit by the Sun's rays. Indeed this would be a correct answer if we could demonstrate that the contribution to sky brightness by other luminous objects in the Universe is negligible. What Olbers found was in fact the opposite! His somewhat startling conclusion was based on the simple calculation which is given below.

Olbers assumed the Universe to be infinitely old and infinite in extent. He further assumed that it always has a uniform distribution of identical luminous objects; and he then proceeded to calculate the amount of light received at the Earth from all these objects.

Suppose there are n such objects in a unit volume and that each object radiates L units of energy in unit time. To simplify the calculation (see Fig. 1.5) we divide space into a large number of thin

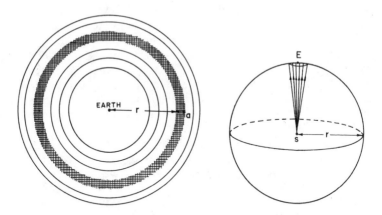

Fig. 1.5.   Fig. 1.6.   s represents the source;
                                              E represents the Earth.

spherical shells centred on the Earth. Let each shell have a thickness a.
Consider a typical shell of inner radius r and outer radius $r+a$. If a
is small, we can approximate the volume of the shell by

$$V = 4\pi r^2 a. \qquad (1.4)$$

Therefore the number of luminous sources in this volume is given by

$$N = nV = 4\pi r^2 an. \qquad (1.5)$$

Since a is small, we can say that a source in this shell is located at a
distance r from the Earth. How much energy does it send across a
unit area at the surface of the Earth held perpendicular to the line of
sight to the source? This can be easily calculated. The energy L
emitted by a source in unit time crosses evenly the surface of a sphere
of radius r centred at the source and passing through the Earth (see
Fig. 1.6). The surface of this sphere has area $4\pi r^2$. Therefore the
amount of energy crossing unit area of this spherical surface per
unit time is

$$l = \frac{L}{4\pi r^2}. \qquad (1.6)$$

The total energy from all sources located in the shell and crossing
unit area at the Earth in unit time is accordingly given by

$$Nl = anL. \qquad (1.7)$$

To get the total energy from all the objects in the Universe we have to add the contributions like (1.7) from all such shells. However, notice that (1.7) gives an answer which does not depend on r, the inner radius of the shell. Thus a remote shell contributes just as much as a nearby one. And since in an infinite Universe the number of such shells is infinite, the total energy received at the Earth across unit area in unit time is also infinite. Luminous objects in the Universe therefore contribute so much light that the sky should be infinitely bright.

This was the conclusion arrived at by Olbers. Notice that for this conclusion to be valid a typical luminous object need not be very bright. The vastness of the Universe takes care that the sky is infinitely bright whether or not we are facing towards or away from the Sun.

This drastic conclusion can be tempered somewhat by arguing that the luminous objects involved are not point sources but have finite dimensions. Thus the light from very remote objects is blocked by other objects lying in its way. A fresh calculation taking this blocking effect into account can be made. The result is that the sky brightness should match the surface brightness of a typical object. Now if the typical object is as bright as the Sun, the temperature in our neighbourhood should be as high as that found on the surface of the Sun, that is, close to 5800 K.† So even this conclusion is not quite satisfactory!

Clearly something must be wrong with the assumptions made by Olbers. Several attempts were made by succeeding physicists and astronomers to modify these assumptions so as to arrive at the result that the sky should be dark at night. The paradox, however, was not resolved until more than a hundred years later, when more detailed information on the structure of the Universe became available. I will postpone discussing the resolution of the Olbers paradox to Chapter 6. I present it here as a good example of how a routine question relating to the local environment (in this case the appearance of the sky) can force a physicist to think of the Universe as a whole.

Can we say that there are situations which show that a secret of the Universe as a whole is concealed in a local phenomenon? In Chapters 5 and 6 I propose to examine this question of the relevance of the

---

† K stands for Kelvin, the unit of temperature in the International system of units; a temperature interval of 1 K is the same as a temperature interval of 1 °C, but 0 °C = 273 K.

Universe to local physical laws. In contrast to the earlier chapters of this book, which represent the impact of laboratory physics on astronomy, these later chapters will try to make a case for the reverse effect. In other words, while the laboratory physicist has contributed greatly towards man's understanding of the structure of the Universe, can he afford to isolate himself from the flow of information available from the Universe at large?

2

The Life of a Star

The confidence felt by an astronomer in the correctness of his theory was never better expressed than by Eddington when defending his theory of stellar structure. Eddington had argued that the central region of a star is hot enough to sustain thermonuclear reactions. When atomic physicists expressed their doubts, Eddington disagreed. Writing in his book *The internal constitution of stars*,† he had this to say: 'We do not argue with the critic who urges that the stars are not hot enough for this process; we tell him to go and find *a hotter place*.' Subsequent developments in nuclear physics have shown that Eddington was right.

Anyone looking at the sky on a clear night is impressed by the enormous number of stars everywhere. While it is relatively easy to conclude that all these stars are similar objects, it is not so obvious that the Sun also is one of their class. After the time of Copernicus, man became aware of the dominating role of the Sun in his planetary system. But before Copernicus, indeed even going back to the primitive days of astronomy, the Sun had already created a special place for itself in the minds of people on Earth simply because it outshone all other objects in the Universe. Yet however important the Sun may be as seen from the Earth, or from any other planet of the Solar System, it is just one of the hundred thousand million stars in our Galaxy. In this chapter we shall look at some of the important characteristics of a star and how they change as the star evolves with time. Thanks to the application of modern laboratory science, it is now possible to answer many questions about stars which puzzled the astronomers of 50 years ago.

Stars are of various types, ranging from stars with a fraction of the solar mass to those 10–100 times more massive than the Sun. There

† See p. 301, First Edition (1926), Cambridge University Press.

are stars with differing chemical composition, different temperatures, different sizes, and so on. While such a variety of stars may look bewildering at first, it helps the astronomer in understanding many important fundamental questions about the Universe, for example: How does a star change its size and brightness as it gets older? How does the chemical composition inside it change? Why do some stars explode? Is there such a thing as the death of a star?

If we ask the question, 'How does the Sun change as it becomes older?', the direct answer would be to wait and watch it. Unfortunately, the Sun does not seem to change perceptibly during a human lifespan, or indeed over several centuries. So how do we answer the question? To understand the way astronomers tackle this problem, let us first consider an analogous problem. Suppose someone from outer space wants to know what happens to a human being over its entire lifespan. If there were only one human being available for study, the only method would be to watch the human being grow from a baby to an adult and then gradually see him become old. This takes several decades. If, however, an entire human population is available for survey, the observer from outer space can form a fairly correct impression of the lifespan of a human being by observing humans of different ages—provided, of course, that he has a good theory of biological development. Unlike the previous case, the observations are made over a whole population instead of on one specimen. Moreover, the time of observation is very short—it need not cover the lifespan of a human being.

The astronomer is in this latter situation when it comes to studying the life of a star. He has stars at various stages of their life on view, and from this picture he has to build up a consistent picture of how each star evolves through its various stages. For this he needs a good theory of stellar evolution. So the question 'What will happen to the Sun?' can be answered by looking at other stars which are judged to be older than the Sun. How is such a judgement made?

The classification of stars

If we pursue further our analogy of the observer from the outer space looking at the human population, the observer would try to classify humans according to their physical characteristics (height, weight, colour, etc.). He would then try to build up a picture of how ageing process changes these characteristics—for example, hair turning from dark to white. The astronomer does the same with stars—trying to see how their characteristics change with them. What are these characteristics? Before coming to a discussion of the characteristics of a star it is necessary to say something about an important factor which affects their measurement. This is the distance of the star from us.

Measurement of stellar distances

The commonest method of measuring the distance away of a star is based on the principle of 'parallax', that is, the change in the apparent direction of a star when seen from different points on the Earth's orbit round the Sun.

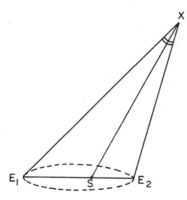

FIG. 2.1. Parallax of star $X = \frac{1}{2}$ the angle E_1XE_2.

Fig. 2.1 shows a star at X and the Earth at two extreme positions E_1, E_2 when it is farthest to one side or the other of the line joining the star X to the Sun S. The directions E_1X and E_2X are different and the angle E_1XE_2 can be measured. In general this is very small and has to be measured with great care. Knowing the positions E_1, E_2 and data on the Earth's orbit, it is possible to compute the distance of the star from the Sun or from the Earth at any time of the year. Half the angle E_1XE_2 is called the parallax of the star.

The distance at which a star would have a *par*allax of one *sec*ond of arc (= 3600th part of a degree) is called the *parsec*. The parsec (pc) and the kiloparsec (kpc) (= a thousand parsecs) are the units used for measurements of stellar distance in our Galaxy, while the unit megaparsec (Mpc) (= a million parsecs) is more suitable for extra-galactic distances.

Because the astronomer is constrained to work from the Earth (or its neighbourhood if he uses modern space technology) he has to make due allowance for distance when studying any star. Fortunately the distance measurements within the Galaxy are relatively free from ambiguity—which is unfortunately not true for extragalactic objects. The accurate measurements of Galactic stellar distances has enabled the astronomer to form a reasonably accurate picture of various stellar properties. The important ones of these are described below.

1. *Brightness*. Stars vary in their brightness considerably. How-ever, we must distinguish between the intrinsic brightness (or 'luminosity') of the star and the brightness as seen from the Earth. Two stars which are equally bright intrinsically may look different to us because of their distance away: the more distant of the two would look fainter. Provided we know the distance of the stars we can make a clear distinction between the two brightnesses.

To measure these brightnesses the astronomer uses the so-called 'magnitude' scale. The 'absolute magnitude' measures the intrinsic brightness of the object, while the 'apparent magnitude' refers to the brightness as seen from the Earth. The scale of magnitudes is a peculiar one, but has been in use for historical reasons. It is based on 'Fechner's law', which says that differences in brightness which correspond to the same fractional part of the whole are equally perceptible by the human eyes, whether the whole intensity is great or small. The 'magnitude' is therefore designed to quantify the factor by which one object appears brighter than another. It is such that a magnitude corresponds to a brightness ratio of about 2·512. More exactly, 5 magnitudes' difference in two stars implies a ratio of 100 to 1 in brightness. Thus if we have six objects A,B,C,D,E,F with the property that A is one magnitude brighter than B, B one magni-tude brighter than C, and so on, then we say that A is 2·512 times brighter than B, B is 2·512 times brighter than C, and so on. Thus A is $2·512 \times 2·512 \times 2·512 \times 2·512 \times 2·512 = 100$ times brighter than F (the number 2·512 is the fifth root of 100). Apparent magnitudes are measured in decreasing order of apparent brightness, while absolute magnitudes are measured in decreasing order of intrinsic brightness (see Box 2.1 for details). Thus if we have two objects A and B with A

---

†Box 2.1 Stellar magnitudes

Suppose a star has a luminosity L, and is located at a distance r from us. L measures the amount of energy emitted by the star in unit time and is a measure of its intrinsic brightness. On the magnitude scale it is measured by the *absolute magnitude M*, which is given by

$$M = -2{\cdot}5 \log L + \text{constant}. \tag{1}$$

The constant is so chosen that $M = 0$ for a star with luminosity $L \cong 3 \times 10^{28}$ W (W = watts). The definition (1) shows that if L increases by a factor 100, M *is reduced* by 5.

We saw in Chapter 1 that a unit area at the Earth held perpendicular to the line of sight to the star will receive an amount of energy

$$l = \frac{L}{4\pi r^2} \tag{2}$$

in unit time. The *apparent magnitude m* measures l on a logarithmic scale. It is given by

$$m = -2{\cdot}5 \log l + \text{constant}. \tag{3}$$

The constant is so chosen that $m = 0$ for an amount of energy flux $2{\cdot}52 \times 10^{-8}$ W m^{-2} outside the Earth's atmosphere.

A comparison of (1) and (3) will show that M is the apparent magnitude of a star if it is measured at a distance of 10 pc without absorption (that is, assuming the inverse-square law (2) to hold).

These magnitude definitions refer to the total energy output from the star and are called the 'bolometric magnitudes'. In practice the measurements relate to specific ranges of wavelengths of radiation. Magnitudes measured in such ranges are named differently; visual magnitudes, photographic magnitudes, etc. The wavelength range in each case is specified.

---

intrinsically 100 times brighter than B, then the absolute magnitude of B will be 5 units more than the absolute magnitude of A. If A and B are two objects such that B appears 100 times fainter than A *as seen from the Earth*, then B has an apparent magnitude 5 units greater than the apparent magnitude of A. Clearly, in the second example, if A and B are equally bright intrinsically, then B must be further away than A. Thus for objects of the same intrinsic brightness the apparent magnitude *increases* with distance.

2. *Colour.* In the above description of a star's appearance, we have omitted one important characteristic—the colour. When we

† Boxes marked with this sign are of a more technical nature and require some knowledge of mathematics and physics, often to undergraduate level.

18 THE STRUCTURE OF THE UNIVERSE

measure the brightness of an object, we in fact measure the energy
received from the object per second, over a unit area, say 1 m². In
practice this energy is distributed over different wavelengths (see
Box 2.2), and our measuring apparatus may respond to only a

Box 2.2 Colours and wavelengths

One of the most significant discoveries of physics in the last century
was that light is in fact a propagation of electromagnetic disturbances
in the form of a wave. A typical light ray consists of periodically
changing electric and magnetic disturbances in directions transverse
to the direction of propagation of light. The frequency of these
disturbances is the frequency of light, and is often denoted by the
symbol ν. This means that in unit time ν oscillations of the electromagnetic field take place, and in this time the light wave advances a
distance c, the velocity of light. The distance travelled in one oscillation period is called the wavelength and is denoted by λ. Thus we have

$$\lambda = c/\nu.$$

In its simplest form light consists of waves of a single frequency (or
wavelength). The human eye is capable of responding to a limited
range of wavelengths—approximately 4000 Å to 8000 Å (Å = ang-
strom, see footnote to p. 19). In this so-called visual range, the
different wavelengths correspond to different shades of colour. The
following table gives a broad distribution of these with regard to the
familiar colours. It is well known that ordinary sunlight is made up of
several colours, that is, it is a superposition of several wavelengths.

Colour	Wavelength (Å)
Violet	3900–4550
Blue	4550–4920
Green	4920–5770
Yellow	5770–5970
Orange	5970–6220
Red	6220–7700

What about electromagnetic waves beyond the visual range? These
will be considered in Chapter 3.

---

limited range of these. So when we measure the brightness of a star
we have to specify over what range of wavelengths this is measured.
For example, the human eye is more sensitive to red light than to
blue light, whereas a photographic plate reacts to these colours in the
opposite way. So a red star may look brighter than a blue star to
the naked eye and yet produce a fainter image on a photographic

THE LIFE OF A STAR 19

plate. Moreover, by using a suitable colour filter the astronomer often measures the light output from a star in a specific wavelength range.

For these reasons, the 'magnitudes' described above have to be characterized by wavelengths or colours corresponding to the method of observation. Hence we have the 'visual magnitude' which is standardized to the wavelength 5500 Å, the 'photographic magnitude' which is characterized by the wavelength 4500 Å, and so on.* Suppose a star is observed at these two wavelengths and it has different apparent photographic and visual magnitudes. If the former is greater, we say that the star is 'red': if the latter is greater, we consider it 'blue'. This relative importance of red or blue is in comparison with a standard star which the astronomers have chosen to be the star Sirius.† The difference between the photographic and

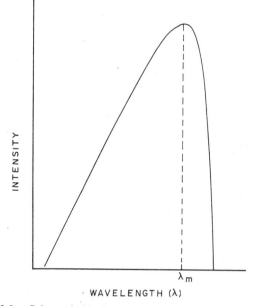

FIG. 2.2. Schematic black-body curve.

* 1 Å = 1 angström = 10^{-10} m. Wavelength is sometimes measured in nanometres (1 nm = 10 Å).
† Also known as the Dog Star, or as Vyadha in India, it is easily the brightest star visible at night.

visual magnitudes is called the 'colour index'. The index is positive for all stars redder than Sirius, and negative for all stars bluer than Sirius.

3. *Temperature.* The colour index provides us with important information about the surface temperature of stars. To understand this let us first consider what is called 'black-body' radiation. This arises, in theory, when we have an ideal enclosure containing sources of radiation. The word 'ideal' means that the radiation is not allowed to escape. It can be shown that such radiation settles down to a definite distribution of energy over different wavelengths.

In Fig. 2.2 we have a typical curve showing how energy is distributed in such a system. We see that it has a maximum at the wavelength $\lambda = \lambda_m$. The shape of the curve depends only on the temperature of the 'black body'. The higher the temperature the lower is λ_m

---

†Box 2.3 Black-body radiation

Imagine an enclosure which contains objects capable of emitting and absorbing electromagnetic radiation. Also suppose that the enclosure is such that the radiation is not allowed to escape. Such a system is called a 'black body'.

In a black body the radiation soon attains an equilibrium distribution of intensity. That is, corresponding to any given frequency (or wavelength) there is a specified amount of radiation. If the radiation at that frequency exceeds the specified amount it gets absorbed; if it is less, it is replenished by the sources present. Such a distribution is called black-body radiation.

Theoretical computations have shown that in a black-body distribution of radiation, the amount of energy per unit volume per unit wavelength range near the wavelength λ measured in metres is given by

$$u_\lambda = \frac{a_1}{\lambda^5 \left\{ \exp\left(\dfrac{c_2}{\lambda T}\right) - 1 \right\}}$$

where $a_1 = 5 \times 10^{-24}$ J m, $c_2 = 0.0144$ m K. T denotes the temperature of the system. To understand the role of temperature consider Fig. 2.2, where u_λ is plotted against λ. The curve has a peak at the wavelength

$$\lambda_m = \frac{0.0029 \text{ m}}{T}. \tag{1}$$

Thus the radiation is dominated by that at wavelength λ_m, and this is related to the temperature by (1). The higher the temperature the lower the wavelength. Thus the temperature of a blue star is higher than the temperature of a red star.

The *total* energy per unit volume is given by summing u_λ over all wavelengths. The answer is

$$u = aT^4,$$

where

$$a = 7\cdot5 \times 10^{-16} \text{ J m}^{-3} \text{ K}^{-4}.$$

Thus the amount of radiation which can be accommodated in a given enclosure rises steeply with temperature.

Ideally, a black body is totally sealed from outside, like a closed, heated oven, and so its properties can only be studied from within. If, however, a tiny fraction of the radiation is allowed to escape outwards, it can carry information about black-body conditions to an outside observer. Clearly, for the black-body character to be preserved inside, the leakage of radiation must be small, and this condition is usually satisfied in the case of a star. For this reason, stars are treated as approximate black-body radiators.

(see Box 2.3). In fact as we vary the temperature T, the product $T\lambda_m$ stays constant.

Now suppose we regard stars as black bodies. Then for each star the temperature at the surface will determine the wavelength at which most of its energy is radiated. And it is this wavelength which will decide the colour of the star. Therefore observation of the colour index tells us about the surface temperature of the star. To visualize this we may take the analogy of heating a block of iron. As its temperature rises so its colour changes from red to blue.

Naturally, all this depends on the stars being like black bodies. This is not exactly right in practice, although the approximation is a good one for all but the very cool stars. For the Sun, the surface temperature determined in this way is about 5800 K.

4. *Spectrum.* Temperatures of several thousand degrees give rise to excitations of the atoms and molecules in the stars and produce spectral lines, which can be observed with the spectrographs attached to the telescopes. A typical star spectrum is shown in Fig. 2.3.

What does a spectrum like this tell us? First, the wavelengths of the lines can be measured and compared with the list of wavelengths compiled by physicists in the laboratory. Spectral lines can be of two types—the 'emission' type (bright bands) and the 'absorption' type (dark bands). The former correspond to radiation emitted by excited

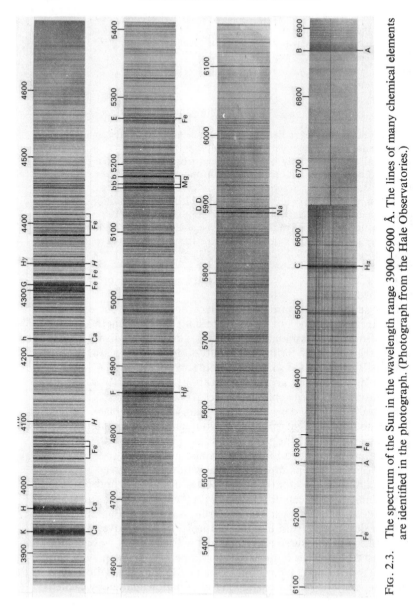

Fig. 2.3.   The spectrum of the Sun in the wavelength range 3900–6900 Å. The lines of many chemical elements are identified in the photograph. (Photograph from the Hale Observatories.)

atoms, while the latter represent absorption of the radiation by cooler material in the intervening space. In either case we get a clue as to the nature of the atoms involved and also as to their relative abundance (see Box 2.4 for details).

Box 2.4 The spectrum of a star

The light from a star, or any other luminous astronomical source, consists of a superposition of electromagnetic waves of several different wavelengths (see Box 2.2). The spectrum of the light source gives a distribution of energy over different wavelengths. A typical spectrum, shown in Fig. 2.3, consists of three parts: (1) the continuum, (2) the emission lines, and (3) the absorption lines.

The continuum represents a continuous distribution of energy over different wavelengths, for example, as seen in Fig. 2.2 for a black body. Superposed on this are the two types of spectral lines, which arise in the following way.

In an atom (or a molecule) the outer part is occupied by electrons. Quantum physics tells us that these electrons can occupy a certain discrete set of trajectories. In the hydrogen atom, for example, the trajectories are orbits round the nucleus, with the electron occupying an outer orbit for larger total energy. When the electron is in the innermost orbit, the hydrogen is in the 'ground state'. In general the ground state represents the state of the lowest energy, and a state of higher energy is called an 'excited state'.

It is possible for the atom or molecule to change its state by absorbing or emitting electromagnetic radiation. If a state of energy E_1 is changed to a state of *lower* energy E_2, the transition is called a 'downward transition', and radiation of frequency given by

$$\nu = \frac{(E_1 - E_2)}{h}$$

is emitted. Here h is Planck's constant. The reverse transition is an 'upward transition', resulting in the absorption of the radiation of frequency ν.

In a star, an emission line in the spectrum arises when a downward transition occurs, leading to the emission of energy. As given by the above formula, the radiation is concentrated in a narrow band of frequencies around ν, and this shows up as a bright line in the spectrum. An absorption line arises when an upward transition occurs, leading to the absorption of radiation from the star in a narrow band of frequencies around ν. This results in the appearance of a dark line in the spectrum.

Stars have been classified according to their spectra in various classes (spectral classes) labelled as O,B,A,F,G,K, and M (remembered easily through the sentence 'Oh Be A Fine Girl, Kiss Me'). This sequence is a temperature sequence with the cooler stars coming later. The class O contains lines of ionized helium, oxygen, nitrogen, carbon, silicon, and so on. The class K contains neutral metals and some molecular bands. Such information, as we shall soon see, plays a vital part in placing the stars in appropriate places on their evolutionary track.

5. *Size*. The Sun has a diameter of about 1·4 million km. The Sun is neither the smallest nor the largest star in the Galaxy. There are stars much larger than the Sun in linear size—some have a radius larger than that of the Earth's orbit. These are called 'giants' and 'supergiants'. Then there are very small stars with diameters no greater than those of, say, Jupiter or Saturn. These are called 'dwarfs'. A dwarf can be highly dense. A cubic metre of a dwarf star can contain several thousand tons of matter.

What place do such stars have in the life-track of a typical star? Does every star go through a 'giant' phase and a 'dwarf' phase in its life? In particular, will the Sun become a giant star and swallow up the Earth and the inner planets? These questions can now be answered with some confidence by present-day astronomers.

The Hertzsprung–Russell diagram

Returning now to our analogy with the human population, suppose that the observer from outer space decides to plot two characteristics of the humans he finds on Earth on graph paper: their weights along the x-axis and their heights along the y-axis. For a representative section, for example, for a population from a residential suburb of a city, the plot may look like that shown in Fig. 2.4. Over the section AB both height and weight appear to increase, although this section contains not as many points as the next section BC, which is more or less a plateau. If he has a sufficiently accurate biological theory, the observer may be able to conclude that these two sections correspond to childhood and to adulthood respectively. Fewer points in the first phase than in the second indicate a shorter time scale for childhood, when the human being is growing, than for middle age. The plateau indicates the limiting height attained by a typical human being and maintained throughout the middle age. Naturally, information on other human characteristics would help in building a more complete picture of human development.

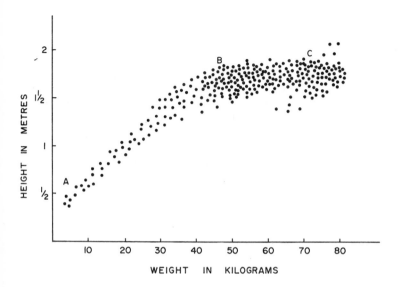

FIG. 2.4.

In the same way, in the study of stellar evolution the Hertzsprung–Russell diagram (known briefly as the H–R diagram) plays a very important part. An H–R diagram has the spectral class of a star plotted on the x-axis and its absolute magnitude on the y-axis. Because of the connection between spectral class and temperature, and between absolute magnitude and luminosity, this is the same as plotting the luminosity of the stars against their surface temperatures.

A typical H–R diagram for the nearest stars is shown in Fig. 2.5. The diagram shows immediately that stars seem to be distributed in a pattern. A majority of the stars are along a curved line BR, with the blue stars at the top end B and the red stars at the bottom end R. This sequence of stars is called the 'main sequence'. Some stars branch off to the right from the main sequence. These turn out to be the giant stars. The dwarfs lie along another line below the main sequence.

As we shall see later, the H–R diagram suggests some sort of connection between stars of different types. Does a typical star move along the H–R diagram in a certain well-determined pattern as it

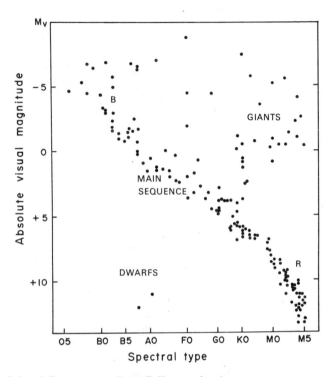

FIG. 2.5.   A Hertzsprung–Russell diagram for the nearest stars.

grows older? In that case a population of stars of various ages would appear to fill that pattern. This is the basic clue to follow in understanding the life of a star.

The birth of a star

How are stars formed in the first place? Although this question has been the subject of research for a good many years, we still do not have a definite answer. However, a somewhat tentative suggestion can be made, and it goes something like this.

It is believed that stars condense out of clouds of gas and dust found in the Galaxy. Typical densities in such regions are around

10^{-19} kg m^{-3}, that is, about a hundred million (10^8) hydrogen atoms per cubic metre. These regions are in fact predominantly made of neutral hydrogen and are called the HI regions. Their temperatures are about 100 K (about 173 °C *below* the freezing point of water). A cloud of gas large enough to contain as much matter as there is in the Sun will not have a gravitational field strong enough to induce contraction. However, if we take a cloud say 100 times greater or more, its gravitational pull will be strong enough for this purpose (see p. 149). Such a cloud contracts as a whole, but subsequently breaks up into smaller subunits or 'protostars' when instability develops in the system. This process of fragmentation, suggested by Sir Fred Hoyle about 20 years ago, leads to the formation of clusters of stars rather than isolated single stars.

There are, however, some difficulties with the condensation idea. The main one is presented by the angular momentum (see Box 2.5),

---

Box 2.5 The angular momentum

Imagine a particle of mass m going round in a circle of radius r with velocity v. At any moment (see Fig. 2.6) the velocity of the particle is in a tangential direction, and in that direction it has a linear momentum given by

$$P = mv.$$

The particle also possesses angular momentum about the axis through the centre of the circle and perpendicular to it, amounting to

$$M = mvr.$$

A more general definition of the angular momentum of an extended object (instead of the point particle discussed above) about *any* axis can be given. The laws of dynamics (based on Newton's laws of motion) tell us that the angular momentum of an object about a fixed axis in space, under no external forces, remains constant. During its overall motion the object may change its shape, the speed of rotation of its different parts, its density, etc., but in such a way as to keep the total angular momentum constant. This powerful result is very useful in discussing the dynamical evolution of astronomical objects like stars and galaxies.

---

and can be described as follows. The cloud of gas contains particles moving at random, thus generating turbulence. This leads to the cloud as a whole possessing an overall angular momentum. Indeed, it would be surprising if the angular momentum were to be zero. Now as the cloud contracts the angular momentum stays constant,

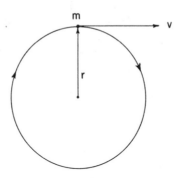

Fig. 2.6.

but its overall rotation increases. (This effect is based on the same dynamical principle which enables ice-skating experts to reduce or increase their speed of revolution by flinging out their hands or drawing them in.) Since the contraction is by a factor of 10^8 or so in linear size, the corresponding increase in rotational velocity is by a similar factor. Stars, on the other hand, do not appear to show such rapid rotation, thus indicating that, somewhere on the way to a star's formation, angular momentum has been lost.

It has been suggested in the case of the Sun that planets take away most of the angular momentum from the parent body and that the Sun's magnetic field provides a way of achieving this. The result is that the Sun itself is left with a small rotational velocity. It may well be that other stars do the same; that is they get rid of their large angular momentum by forming planetary systems. It is also possible that the angular momentum of the original gas cloud is converted into the rotational velocities of the stars as they move in a cluster. Thus a typical star would not revolve round some axis but rather would move round the cluster in which it was situated.

These difficulties are not insurmountable; they will, no doubt, be resolved when a better understanding has been reached of the interstellar medium and of the operation of electromagnetic forces in fluid media. For the present purpose though we shall assume that condensation into a star-like object with very little angular momentum has taken place. The subsequent behaviour of the object can then be followed with less difficulty.

The next stage consists of the transformation of the condensed object into a star which emits light. This transformation occurs rapidly. For, in the initial stages, the force of gravitation between the different constituents of the object is so strong as to cause a rapid collapse of the object as a whole. This, however, does not occur smoothly. Rather, the process is somewhat catastrophic, generating shock waves throughout the star. Heat is generated, owing to the compression. These effects result in the build-up of strong internal pressures in the star, which tend to slow down the compression— until the object settles down to a more-or-less static state. In this state the pressure forces are in balance with the gravitational force. If such pressures were not built up a star like the Sun would have collapsed drastically within less than half an hour. Even when pressures are present, the balance between the two opposing forces has to be exact, otherwise the object soon develops dynamical motions. The heat generated during compression is radiated by the object, which can now be called a 'star'.

In the early stages of its evolution the star undergoes changes in its luminosity while it is in 'hydrostatic equilibrium', that is, while it is being held in balance between the pressure force and the gravitational force. These early stages can be described on the H–R diagram by the so-called 'Hayashi track' (see Fig. 2.7). The star as it evolves moves vertically down the graph, thus indicating a decrease in luminosity (L), and then turns left, implying a rise in surface temperature (T). How and when does this take place? These questions have been answered by Hayashi, Iben, and others with detailed mathematical calculations. Here I shall only summarize their findings.

To start with, the star is wholly 'convective'. That is, heat energy is transported from its interior to the outer regions of the star by the actual transport of fluid, much in the same way that heat is carried when we boil water in a kettle. As the star gradually contracts, its luminosity decreases until it falls below a critical value. This value represents the minimum luminosity possible in a completely convective star. The star therefore is no longer completely convective, and it develops a 'radiative core'. That is, the energy transport in the core takes place by radiation rather than convection. This means that there is no longer any transport of material within the core, and this naturally alters the equations determining the structure of the star. The star's evolutionary track is changed. The track turns leftward until it reaches another critical stage. This corresponds to a rise in the

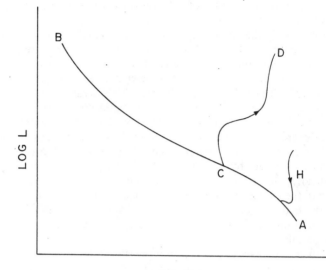

FIG. 2.7. A schematic type of Hertzsprung–Russell diagram plotted using log L versus log T, showing a Hayashi track.

central temperature of the star so that it becomes a 'thermonuclear reactor', thus leading to the next stage in the star's life.

The main sequence

'What keeps the Sun shining?' The question has been there ever since man first looked at the heavens, but has been answered with some measure of confidence only in the last three or four decades. Clearly some reservoir in the interior of the Sun must supply the energy which it continuously radiates. Without any scientific background, primitive man naturally ascribed the brightness of the Sun to some divine power. And even the scientist fared no better, until the present century, in locating this mysterious energy reservoir. In the nineteenth century, two great scientists, Kelvin and Helmholtz, suggested that the energy reservoir is gravitational in origin, that is, the Sun is slowly contracting and therefore losing gravitational energy. This loss of energy appears as the heat and light radiated by the Sun.

However, the Kelvin–Helmholtz contraction hypothesis implied that the total energy available in this way from the time the Sun was born to its present state would have been spent within a few tens of millions of years (see Box 2.6 for details). In other words, the Sun

---

†Box 2.6 The Kelvin-Helmholtz contraction

A uniform sphere of mass M and radius R possesses a gravitational energy given by

$$V = -\frac{3}{5}\frac{GM^2}{R},\qquad(1)$$

where G is the gravitational constant. Suppose such a sphere slowly contracts. From (1) we see that a decrease of R means a decrease in V. In particular, if we imagine an object of the mass of the Sun shrinking from infinite radius to the present radius $R_\odot$ of the Sun, the above formula tells us that the total energy released is

$$\mathscr{E} = \frac{3}{5}\frac{GM^2}{R_\odot}.\qquad(2)$$

The Sun is not actually a homogeneous sphere; but allowing for this requires only a modification of the factor $\frac{3}{5}$ in (2) to another factor which is not markedly different. So we will keep the factor $\frac{3}{5}$ in our further discussion. For the Sun, we can substitute the known values of $M_\odot$ and $R_\odot$ to get

$$\mathscr{E} = 2\cdot4\times10^{41}\text{ J}.\qquad(3)$$

The Sun is emitting radiation at the rate of

$$L_\odot = 4\times10^{26}\text{ J s}^{-1}.$$

If the Sun has been radiating at this rate since its birth, that is, since the time when its initial radius was very large, it cannot be older than

$$t = \frac{\mathscr{E}}{L_\odot} \simeq 6\times10^{14}\text{ s},$$

that is, about 20 million years old.

In the last century Kelvin and Helmholtz employed arguments like the one above to explain the radiation from the Sun. But we see that, if the radiation arises entirely at the expense of gravitational energy, the Sun cannot be more than a few million years old. This, as is explained in the text, is too low an estimate for the Sun's age.

---

cannot be older than, say, 20 million years or so. But this is too small an age. The age of the Earth itself is estimated by geophysicists to be about 4000–5000 million years, and we do not expect the Sun to be younger than the Earth. Therefore gravitational energy cannot be the main source of Sun's brightness.

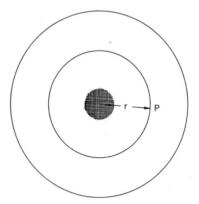

FIG. 2.8.   Schematic diagram for stellar interior. The shaded region represents the central part where energy generation takes place.

In the late 1920s the Cambridge astronomer Eddington made remarkable progress in writing down the equations of stellar structure. These equations (see Box 2.7) describe basically how a star is held in

---

†Box 2.7 The equations of stellar structure

The basic equations describing the equilibrium of a star and the behaviour of matter inside it are as follows. Suppose the star has an overall mass M and radius R. Consider an interior point P at a distance r from the centre of the star. Draw a sphere of radius r concentric with the surface of the star (see Fig. 2.8) with the point P on it. A piece of matter at P is attracted inward, owing to the gravitational pull of the rest of the matter in the star. The Newtonian law of gravitation tells us that this pull is towards the centre and is equal to that exerted by the mass $M(r)$ interior to the sphere of radius r. For this matter to remain stationary at P this pull must be countered by the force due to the pressure p in the star. This balance is described by the equation

$$\frac{\mathrm{d}p}{\mathrm{d}r} = -\frac{GM(r)\rho}{r^2}, \tag{1}$$

in which ρ is the density of matter at P. dp/dr is the pressure gradient which provides the force. Since the right-hand side of (1) is negative, the pressure decreases outwards. It vanishes at the surface of the star (or equals the pressure of the atmosphere of the star, if it exists).

The second equation describes the geometrical relation of mass to density:

$$\frac{dM(r)}{dr} = 4\pi r^2 \rho. \tag{2}$$

The next equation, called the equation of state, relates the pressure to the density and temperature T:

$$p = \frac{\mathscr{R}}{\mu}\rho T + \frac{1}{3} aT^4. \tag{3}$$

Here the first term on the right-hand side describes the pressure of the gaseous material in the star, while the second term describes the pressure due to radiation. $\mathscr{R}$ and a are constants which can be derived from laboratory physics experiments. μ is the mean molecular weight of the matter at P and depends on the chemical composition of the star: $\mu = \frac{1}{2}$ for a star made up wholly of hydrogen.

The fourth equation describes transport and absorption of energy across the star. If $L(r)$ is the energy flowing outwards from the surface of the sphere of radius r, and K is the opacity of matter at P (that is, a measure of its absorbing power) then the outward transport of energy is accompanied by a drop in temperature as given by

$$\frac{4}{3} aT^3 \frac{dT}{dr} = - \frac{K}{4\pi c} \frac{L(r)}{4\pi r^2} \rho. \tag{4}$$

The temperature may be millions of degrees at the centre, but will drop to a few thousand degrees towards the surface of the star.

Finally, if ϵ is the rate of energy generation per unit volume, we have

$$\frac{dL(r)}{dr} = 4\pi r^2 \epsilon. \tag{5}$$

ϵ is determined from the data on relevant thermonuclear reactions.

Even crude approximation of these equations yields interesting information. Suppose P_c and T_c are the pressure and temperature of the Sun at the centre. We will approximate dp/dr by $-P_c/R$, and get from (1) the following rough estimate:

$$\frac{P_c}{R_\odot} = \frac{GM_\odot}{R^2_\odot} \rho.$$

Next we will ignore the radiation pressure in (3) and write

$$P_c = \frac{\mathscr{R}}{\mu} \rho T_c.$$

Substituting this value of P_c in the first equation we get

$$T_c = \frac{\mu}{\mathscr{R}} \cdot \frac{GM_\odot}{R_\odot}$$

With $\mu = \frac{1}{2}$, $\mathscr{R} = 8.3$ J K^{-1} per molecule, $R_\odot = 7 \times 10^8$ m, and $M_\odot = 2 \times 10^{30}$ kg, we get T_c as nearly 10 million K. (The more exact calculation gives T_c as nearly 13 million K).

hydrostatic equilibrium and how the transport of energy from the inner regions to the outer regions takes place in a star. The equations form an interlinked set, and can best be solved using an electronic computer. The astronomer specifies the total mass of the star and its chemical composition, and the solution of the equations then determines all other parameters, for example, the radius, the central temperature, and the light output.

However, this remarkable set of equations lacked one important piece of information—the rate of energy generation within the star. In the late 1920s and early 1930s Eddington had suspected that the source of energy was to be found inside the nuclei of the atoms which make up a star. When nuclei are transformed from one state to another this energy can be released in some cases. However, the atomic physicists of those days did not think that such a release of energy was very feasible. Their main objection was as follows. For nuclear transformation to occur, two or more nuclei of like charges have to be brought together. However, like charges repel one another, and in order to surmount this repulsion the charges are required to move towards one another with tremendous velocity. This is possible only if the temperature of stellar material is very high. How high? Eddington estimated the central temperatures in stars to be as high as 40 million K, and confidently believed that energy-producing nuclear reactions must be triggered off at such temperatures. The atomic physicists disagreed. They felt that these temperatures were too low for the purpose of releasing nuclear energy.

A few years later Eddington was proved right. In 1939 work by Hans Bethe and others showed how light nuclei can be brought together and transformed into heavier nuclei at temperatures around 10 million K, which are easily achieved in the centre of a star. The process of nuclear transformation results in the release of energy, which provides the source for the brightness of a star like the Sun. We shall now consider some details.

The process may be described briefly as the fusion of four hydrogen nuclei to form a helium nucleus. Symbolically this may be written

$$4 \,^1\text{H} \rightarrow \,^4\text{He} + \gamma,$$

where ^{1}H on the left-hand side represents a hydrogen nucleus, with the superscript 1 denoting its approximate mass in atomic mass units. ^{4}He similarly represents a helium nucleus with approximately 4 atomic mass units. *The symbol γ is here used to describe the release of energy in the form of radiation (finally in the form of γ-rays). The energy released when four hydrogen atoms are so fused together is 26·72 MeV.† Where does this energy come from? Its origin lies in Einstein's famous formula

$$E = Mc^2,$$

which says that the energy E released when mass M is destroyed is given by M times the square of the velocity of light c. If we take the mass difference of four hydrogen nuclei and one helium nucleus we will get the amount of energy finally released from the above formula. It is this same process which brings about the release of energy in the hydrogen bomb. So we may argue (rightly) that the Sun is kept shining because hydrogen bombs are exploding all over its interior!

In a typical star this process of converting hydrogen to helium takes place in one of two possible ways. The first one, called the proton–proton chain (or p–p chain), operates in stars of small masses (in the range of $1 \cdot 5 M_\odot$–$2 M_\odot$ or less); the second is the carbon–nitrogen–oxygen cycle (CNO cycle), which operates in more massive stars. The net effect of these reactions is the same as that described earlier, namely, the conversion of hydrogen to helium. The details are described in Box 2.8.

Box 2.8 The p–p chain and the CNO cycle

The basic reactions of the p–p chain are as follows:

$$^1\text{H} + \,^1\text{H} \rightarrow \,^2\text{D} + e^+ + \nu$$
$$^2\text{D} + \,^1\text{H} \rightarrow \,^3\text{He} + \gamma$$
$$^3\text{He} + \,^3\text{He} \rightarrow \,^4\text{He} + 2 \,^1\text{H}.$$

That is, first two protons (or hydrogen nuclei ^{1}H) combine to give the

---

* The atomic mass unit (amu) is defined as $\frac{1}{12}$ the mass of the carbon atom of mass number 12.

† MeV stands for million electronvolts, and is equivalent to an energy of about $1 \cdot 6 \times 10^{-13}$ J (J = joule).

deuterium nucleus (^{2}D), with the release of a positron and a neutrino. Next the ^{2}D combines with ^{1}H to form the ^{3}He nucleus, and releases energy. Finally two ^{3}He nuclei combine to form the more stable helium nucleus (^{4}He). In this process two protons are released. A proper accounting of these shows that four ^{1}H nuclei combine to give a ^{4}He nucleus with the release of 26·71 MeV energy. The p–p chain can also proceed along other channels in which the short-lived lithium, beryllium, and boron (^{7}Li, ^{7}Be, ^{8}B) nuclei are formed and destroyed in the intermediate stage.

In the CNO cycle the main set of reactions is given by

$$^{12}C + {}^1H \rightarrow {}^{13}N + \gamma$$
$$^{13}N \rightarrow {}^{13}C + e^+ + \nu$$
$$^{13}C + {}^1H \rightarrow {}^{14}N + \gamma$$
$$^{14}N + {}^1H \rightarrow {}^{15}O + \gamma$$
$$^{15}O \rightarrow {}^{15}N + e^+ + \nu$$
$$^{15}N + {}^1H \rightarrow {}^{12}C + {}^4He.$$

In these reactions the second and the fifth are the slow ones, just as in the p–p chain the first reaction is slow. These involve a change of electric charge. The CNO cycle also has other alternatives besides the one shown here. The energy release is the same as in the p–p chain since basically four ^{1}H nuclei are fused together into a ^{4}He nucleus. The CNO cycle requires a higher operating temperature than the p–p chain, and hence it is more likely to occur inside more massive stars.

These nuclear reactions take a long time (of the order of thousands of millions of years), and while they are going on the star observes this time scale. That is, it remains in a hydrostatic equilibrium for a long time, far longer than the gravitational-collapse time scale described on p. 31. The star's nuclear reactions generate energy which goes towards providing the necessary pressure to support the star against its tendency to contract with gravitation. The long time during which the star is steadily burning its nuclear fuel is called the 'main-sequence' phase in the life of the star. The star lies on the section AB on the H–R diagram (Fig. 2.7)—a line extending from the bottom right-hand end to the top left-hand end. The larger the mass of the star, the higher up it is on the main sequence. Therefore we can roughly divide the main sequence into two parts. The lower main sequence includes low-mass stars which generate energy through the p–p chain, whereas the upper main sequence includes stars of higher mass with the CNO cycle as the process of fusing hydrogen.

The red giant stars

When a star embarks on the main-sequence stage it may start with pure hydrogen as the main or predominant constituent. However, with time, the hydrogen gets converted to helium, from the centre outwards. So the composition of the star becomes inhomogeneous, unless there is some process operating within the star which mixes up all its material. In a completely convective star convection provides such a mixing agent. However, in most other cases we would expect very little mixing throughout the star. This has led theoreticians to look at models of stars with inhomogeneous chemical composition.

The inhomogeneity in composition can be broadly of two types— discrete and continuous. In Fig. 2.9(a) we have an upper-main-

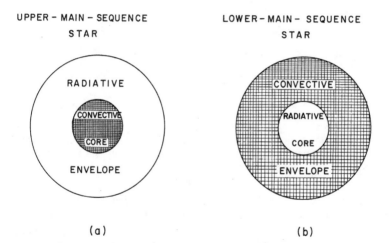

UPPER – MAIN – SEQUENCE STAR

RADIATIVE
CONVECTIVE
CORE
ENVELOPE

(a)

LOWER – MAIN – SEQUENCE STAR

CONVECTIVE
RADIATIVE
CORE
ENVELOPE

(b)

FIG. 2.9.

sequence star using the CNO cycle. In this star it turns out that the interior, or the core, is convective, while the exterior is radiative in character. This means that, as hydrogen is burnt in the interior, there is mixing within the core, so that its composition is uniform but different from that of the exterior. So there is a jump in the composition as we cross the core boundary. In Fig. 2.9(b) we have a lower-main-sequence star. In this star convection operates in the exterior and radiation in the core, and so there is no mixing in the core. This leads to the material within the core becoming gradually less rich in helium as we move outwards from the centre.

The net effect of this inhomogeneity on the structure of the star is spectacular, and more so in the case of the discontinuous change in composition. The overall equilibrium in the star is upset, and in order to readjust it the star has to expand considerably in its exterior region, while its core shrinks because of the superiority of the gravitational forces over the supporting pressure of stellar material (see Fig. 2.10).

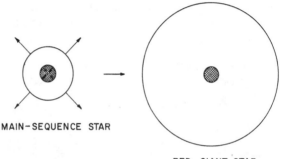

MAIN–SEQUENCE STAR

RED GIANT STAR

FIG. 2.10. Transition from the main-sequence to the red-giant state (not drawn to scale). In this transition the core shrinks while the envelope expands.

Because of its enormous size such a star is called a 'red giant star'. The word 'giant' reflects its size, while the word 'red' indicates its overall colour. The star at C branches off from the main sequence, as shown on the H–R diagram (Fig. 2.7) by the line CD. The evolution to and beyond the red-giant stage involves several intricate details which are still being investigated.

What will happen to the Sun? Like any other star on the main sequence, the Sun (near C in Fig. 2.7) will branch off along the giant track. This will lead to an increase in its luminosity and its radius, but its overall surface temperature will drop. It may become so big that it will swallow the planets Mercury and Venus, and even the Earth. However, many thousand million years will have to elapse before the Sun becomes a red giant.

What are the chemical reactions which sustain a star during its red-giant phase? These are described by the name α-process and involve the build up of nuclei heavier than helium by successive fusion with helium (the ^{4}He nucleus is also called an α-particle). Three helium nuclei combine to give a carbon nucleus ^{12}C:

$$3\ ^4\text{He} \rightarrow {}^{12}\text{C} + \gamma$$

where the presence of γ on the right-hand side indicates that energy is released. The next nucleus to be formed is oxygen (^{16}O), by the addition of another ^{4}He, and so on. I shall return to a description of the synthesis of elements in stars towards the end of this chapter.

The upshot of this is that we have a contracting core in which higher and higher temperatures are developed towards the centre, with the formation of heavier and heavier elements. The composition of the core thus gets progressively more complicated. Detailed calculations are now being performed on modern electronic computers to take these complications into account.

In spite of the involved reactions going on in its interior, the star spends a comparatively short time in the red-giant phase. This is largely because the red-giant reactions do not take as long to complete as the reactions during the main-sequence stage. This time-scale effect shows up on the H–R diagram, where more stars are found on the main sequence than on the giant branch. After completing the red-giant phase the star enters the final stage of its evolution.

White dwarfs, neutron stars, and black holes

The details of how a star behaves after its red-giant stage are not yet fully understood. But we know more about the final state itself than about how the transition to this stage occurs.

For stars with a mass less than about $1 \cdot 4 M_\odot$ (see p. 3), the final state is better understood. $1 \cdot 4 M_\odot$ is known as the 'Chandrasekhar limit', after the work of S. Chandrasekhar, who first obtained this result in 1935. The final state here means the state when the star has exhausted all its nuclear fuel, that is, it can no longer draw on thermonuclear reactions to supply the energy and the pressure needed to withstand gravitation. Such a star contracts until the matter gets highly compressed, so much so that the electrons in it become quantum-mechanically 'degenerate' (see Box 2.9). This degenerate

---

Box 2.9 Degenerate matter

One of the important results of quantum theory is that there exists a class of elementary particles, called 'fermions', with the property that no two identical fermions can be found in the same state. Electrons and protons are fermions.

Now one of the important characteristics specifying the state of a particle is its energy. So in a given system of electrons, the number of electrons occupying a given energy level is limited, because all such electrons must differ in other characteristics which specify the state, for example, the momentum and the spin. The permissible number increases with energy.

In a highly compressed state of electrons (say with a density of 10^9 kg m^{-3}) at low temperatures these energy levels tend to be all filled, and the matter is called degenerate. For such matter the pressure increases with density according to a fixed law. The pressure arises from the above-mentioned result that we cannot squeeze too many electrons of the same energy into a given region of space. The degenerate pressure therefore reflects the resistance provided by matter when an attempt is made to compress the matter beyond a permitted amount. At higher densities, of the order of 10^{15} kg m^{-3}, even neutrons (which are also fermions) become degenerate and behave similarly.

---

matter provides its own pressure, which is now called upon to provide hydrostatic equilibrium. There is, however, a limit to this type of pressure. The bigger the mass of the star, the less effective is this pressure in providing support against gravitational contraction. The Chandrasekhar limit describes this fact. For a star with higher mass than $1 \cdot 4 M_\odot$, the degenerate electron pressure is not effective in providing hydrostatic equilibrium.

Stars with masses less than $1 \cdot 4 M_\odot$, which rely on the degenerate electron pressure, are called 'white dwarfs'. The density of a white dwarf star is very high. Take the case of the white dwarf 40 Eridani B. This has an estimated mass of $0 \cdot 44 M_\odot$ and a radius only $1 \cdot 5$ per cent of that of the Sun. Thus, on an average, a cubic metre of this star contains about 170 000 tonnes (metric tons) of matter. Such stars eventually cool off and become less and less luminous. For example, the bright star Sirius (see footnote p. 19) has a faint white dwarf companion. The pair is often referred to as Sirius A and Sirius B respectively.

In the last few years attention has been focused on another possible star model, which contains matter in an even denser form than that found in white dwarfs. These are the so-called 'neutron stars'. As the name suggests, these stars are made of the elementary particles called neutrons. Normally, a neutron is an unstable particle. In the laboratory a neutron decays into an electron, a proton, and an antineutrino in about 700 s. However, when matter is highly compressed the reverse reaction is more favourable, so that matter made of electrons and protons can combine to give neutrons. For this to occur the matter density must be about 100 times higher than that in a white dwarf. The mass of a neutron star has been estimated to be not higher than about $3M_\odot$.

Although a number of white dwarfs have been detected, no-one has yet observed a neutron star. This may be because neutron stars are very faint—in the visual range a neutron star may be a million or more times fainter than the Sun. However, the detection of 'pulsars' (Box 2.10) in the last 5 years has created great interest in neutron stars. At present reasonably satisfactory models of a pulsar are based on a neutron star. So we may have observed neutron stars already through pulsars.

Box 2.10 Pulsars

The first pulsar was discovered, by the Cambridge radio-astronomers, in 1968. It was an unexpected discovery, but its importance prompted a systematic search for more pulsars by radio-astronomers all over the world and this has led to the discovery of dozens of pulsars. Pulsars are characterized by the extreme regularity and short period (of the order of a second) of their pulses, which sets them apart from other astronomical objects.

Unlike quasars (see p. 84), the pulsars so far found are comparatively nearby. They are in our own Galaxy. Although there are several question marks with regard to their behaviour, it is now believed that pulsars can be associated with rotating neutron stars and that their radiation comes from moving plasma (that is, a system of charged particles) in magnetic fields. The neutron stars may arise (see the text) from the collapsed core of a supernova. So far one pulsar has been detected at the seat of a supernova explosion. This is the Crab pulsar, located in the Crab Nebula (see Box 2.11).

What if the mass of a star is even higher than that of the critical mass for a neutron star? There are two possible courses of evolution open to it. The first is one of continuous gravitational contraction (see p. 149). If we believe Einstein's general theory of relativity implicitly, we have to conclude that the star becomes a 'black hole', that is, it shrinks so much that light from the star cannot get out of its strong inward gravitational pull. Such a star therefore cannot be 'seen', but its presence will be felt through its gravitional field. Thus if the Sun were to become a black hole, the Earth would continue to go round it, but we would not 'see' any object located where the Sun was. So far no black holes have been detected (though the location of a possible one has been suggested); but if any ever are located for certain, their detection will be through their gravitational effect rather than through measurement of their electromagnetic radiation.

The other evolutionary course open to a massive star is more catastrophic, and I will describe it in the following section.

Supernovae

When a massive star has reached the end of its nuclear fuel, it can become a 'supernova'. A supernova arises when the core of a star collapses under its own gravitational attraction, releasing energy which causes the outer envelope to explode. Thus the inner part of the star undergoes an *implosion*, while the outer part undergoes an *explosion* (see Fig. 2.11). The imploding core may form a white dwarf or a neutron star. The exploding envelope carries an enormous quantity of energy outwards, in the form of fast-moving electrons, protons, and nuclei, as well as the electromagnetic radiation.

The actual explosion may last only a short time—say a day or two

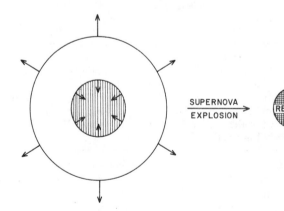

SUPERNOVA
EXPLOSION

REMNANT

FIG. 2.11.

—but its after-effects linger on for much longer. The most spectacular supernova explosion in our Galaxy, as seen from the Earth, is that of the Crab Nebula supernova (see Box 2.11). This supernova has been of interest to the astronomer in various ways. It is a source of light, radio waves, and X-rays, and it also contains a pulsar formed from the collapsed core.

Box 2.11 The Crab Nebula

The Crab Nebula presents the most detailed record of a supernova explosion. Fig. 2.12 is a photograph of the nebula, which now shows the remnant of an explosion which was observed from the Earth on the 4th of July in the year A.D. 1054. This date has been fixed from ancient Chinese and Japanese astronomical records. For example, the following statement appeared in the 'History of Sung Dynasty' (Ho Peng Yoke, 1962), *Vistas in Astronomy* **5**, 184: 'On a Chi-Chhou day in the fifth month of the first year of Chi-Ho reign period a 'guest star' appeared at the south east of Thien-Kaun measuring several inches. After more than a year it faded away.'

This guest star is now believed to be an exploding supernova which must have been very spectacular at the time of explosion. It was so bright as to be easily visible in broad daylight. Recently William C. Miller of the Hale Observatories suggested two possible records of this event in Northern Arizona, one in the White Mesa and the other in the Navaho Canyon. These records consist of drawings by the Pueblo people, one painted, the other incised. The Navaho Canyon drawing is shown in a schematic form, in Fig. 2.13. It shows a crescent in close association with a round object. The crescent no doubt represents the Moon. What about the round object? Miller has analysed the various possibilities and come to the conclusion that it must represent the exploding star now associated with Crab Nebula. The White Mesa pictograph led him to similar conclusions.

The Crab Nebula is of great interest to astronomers because of its many different properties. It emits radio waves, it is a source of X-rays, it is the seat of a pulsar, and it is also a source of high-energy cosmic ray particles.

Supernova explosions have been suggested as sources of cosmic rays, that is, the fast particles from space which eventually find their way into the Earth's atmosphere (see p. 97). These explosions also 'contaminate' interstellar space with the heavier nuclei formed in the star during stellar evolution.

The actual mechanism of the explosion has been studied for a number of years. Theoretical models involving shock waves have been constructed to show how the energy released from the core is transmitted outwards.

FIG. 2.12. (Photograph from the Hale Observatories. (Copyright by California Institute of Technology and Carnegie Institute of Washington.))

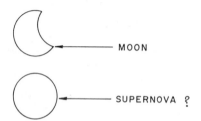

FIG. 2.13. A schematic representation of the Navaho canyon drawing.

The frequency of these explosions in the Galaxy has been variously estimated at around one per 10–30 years, although at present we must treat these estimates with caution. Until we have a more quantitative theory which tells us what proportion of stars ends up as supernovae, we have no real means of estimating this frequency except through direct observations.

A star as a thermonuclear reactor

The interior of a star provides conditions ideal for the fusion of nuclei. Indeed it is not merely the case that nuclear reactions *can* take place in a star but that they *must* take place. The energy released by them provides the pressure which supports the star under its own gravitational contracting force. While nuclear reactions in stars have been considered by many authors, the first comprehensive account of how different elements are synthesized in stars was given by Geoffrey and Margaret Burbidge, William Fowler, and Sir Fred Hoyle in 1957 (these four authors are often referred to as B^2FH in this connection).[†]

The study undertaken by the B^2FH group was partly motivated by their interest in cosmology. 'Cosmology' is the branch of astronomy which deals with the origin and structure of the Universe as a whole. One of the important issues of cosmology, as we shall see in Chapters 4 and 7, is to understand the origin of matter in the Universe. Matter is found in the Universe in various forms, some simple, some complicated. Did it originate in these forms, or did it evolve to complicated structures from relatively simple ones? The cosmologists, by and large, like to believe in the latter possibility. Assuming then that matter was created in a simple form, how did it evolve to the more complicated forms now seen everywhere?

Atomic physics tells us that hydrogen is the simplest element and that its atoms contain a proton at the nucleus and an electron in orbit round it. More complicated atoms contain many protons and neutrons at the nucleus and many electrons in different orbits round the nucleus. Assuming that matter first appeared in the form of neutrons, protons, and electrons, can we understand how the different elements were formed? This is the problem that astrophysicists have been trying to solve.

One important suggestion, which we shall discuss in detail in

[†] The technical account of this work is given in *Rev. mod. Phys.* **29**, 547–650 (1957).

Chapter 4, was made in the 1940s by George Gamow. He assumed that the Universe was created with a big explosion, and that very high temperatures existed in the early stages after the 'big bang'. It was in these few moments after the creation of the Universe that the neutrons, protons, etc., were crammed together to form hydrogen and heavier elements. This process of nucleosynthesis requires temperatures of thousands of millions of degrees and could not operate in a Universe without such a period of high temperatures. For example, it could not operate in a steady-state Universe (see Chapter 4), which does not have high-temperature epochs. What the Burbidges, Fowler, and Hoyle did was to show that there was an alternative source of nucleosynthesis: a star. They pointed out that within a star, during the different phases of its evolution, the central temperature rises steadily until it reaches thousands of millions of degrees. Their detailed work consists of examining the various evolutionary stages of a star to see how the different elements could be manufactured in it. The processes involved are summarized below.

1. *The onset of nucleosynthesis.* Nuclear reactions begin in a star like the Sun when its central temperature reaches about 10 million K (10^7 K). The star, predominantly made up of hydrogen at this stage, begins to act as a nuclear reactor by converting hydrogen to helium. This can happen in two possible ways—either through the p–p chain or through the CNO cycle (see p. 35), depending on the mass of the star.

2. *The helium burning.* During its hydrogen-burning phase the star is largely on the main-sequence track of the H–R diagram. However, as its composition changes, with more and more helium in the core, it begins to branch off along the red-giant track. During the giant stage the core contracts and heats up, while the envelope expands.

When the core temperature reaches 100 million K (10^8 K) and its density reaches about 10^8 kg m^{-3} (that is a hundred thousand times that of water), the 'Coulomb barrier' (the force of electrostatic repulsion between helium nuclei) can be surmounted. At first we might think that two such nuclei might combine to give ^{8}Be (Beryllium 8). However, ^{8}Be is not a stable nucleus, and it breaks up again into two helium nuclei. Indeed to form a stable nucleus we need *three* helium nuclei, which combine to form an excited state of the carbon nucleus which subsequently decays into the ground state:

$$3\ ^4\text{He} \rightarrow {}^{12}\text{C}^* \rightarrow {}^{12}\text{C}.$$

In this reaction the asterisk indicates an excited state of carbon, that is, a state of higher energy than that found in the ordinary carbon nucleus. This energy is only slightly higher than the combined energy of a ^{8}Be nucleus and an α-particle, and this makes helium burning a quick process.

The instability of ^{8}Be and the presence of the excited state of carbon have played a critical role in determining the abundance distribution of the Universe. If ^{8}Be were stable, or if the excited state of carbon had not existed, the Universe would have had a totally different abundance picture, and even life as we know it would not have been possible.*

3. *The α-process.* This is the process whereby heavier nuclei are formed by successive fusion with more and more helium nuclei. These processes take place in giant stars with cores contracting and heating up. Thus we get oxygen (^{16}O), neon (^{20}Ne), magnesium (^{24}Mg), silicon (^{28}Si), sulphur (^{32}S), etc. Notice that the atomic masses are increasing by 4 in this series because of the addition of ^{4}He (that is, an α-particle), which has mass 4.

However, successive addition of an α-particle cannot go on indefinitely, because Coulomb repulsion grows stronger as the nuclear charge increases. Beyond, say, silicon or sulphur, the process operates somewhat differently. Thus one ^{28}Si may disintegrate into seven α-particles, which then combine with another ^{28}Si to form nickel (^{56}Ni). Nickel-56 decays into cobalt (^{56}Co) and iron (^{56}Fe). The process terminates when these three nuclei (nickel, cobalt, iron) of the iron group are formed.

4. *The e-process.* The iron-group nuclei are most stable—in the sense that maximum energy is required to break them apart (see Box 2.12 for a discussion of nuclear binding energies). Towards the end

---

Box 2.12 Nuclear binding energies

Suppose we take a complex atomic nucleus and examine its contents. It will turn out to have a certain number of neutrons and a certain number of protons. If, for example, it has N neutrons and Z protons, we would expect its total mass to be

$$M_0 = Nm_N + Zm_P,$$

where m_N is the mass of a neutron and m_P is the mass of a proton. However, the actual mass M of the nucleus turns out to be different

---

* Compare Sir Fred Hoyle's remarks on this subject in his book *Galaxies, Nuclei and Quasars.* Heinemann, London (1965).

from M_0. The quantity M_0-M is called the 'mass defect' and when converted into energy, using Einstein's relation $E = mc^2$, it is called the 'binding energy':

$$B = (M_0-M)c^2.$$

The binding energy is a measure of work required to break up the nucleus and separate all its constituents until they are far away from one another. This work has to be done against the forces of attraction which exist between nuclear components.

For a stable nucleus the binding energy must be positive. Indeed, the stabler the nucleus, the higher its binding energy. Although the precise nature of the nuclear attractive forces is not known, semi-empirical arguments have been given to arrive at a formula for B in terms of N and Z. From this it turns out that B is highest for the nuclei in the iron group (see Fig. 2.15). Broadly speaking, the nuclear attractive force increases as the number of protons and neutrons is increased. However, if we put in too many protons, their mutual electrostatic repulsion tends to decrease the binding energy. Under these two opposing tendencies, the stablest nuclei are obtained in the iron group. For this reason it is possible to use the α-process until the iron group is reached, but not beyond it.

of the α-process the core of the star has reached a temperature in excess of 1000 million K (10^9 K), while the densities exceed a million times that of water. At this stage a large number of nuclear reactions can occur between the nuclei, the α-particles, the protons, and the neutrons. The best way to treat this problem is by using a statistical method, in which an equilibrium between hundreds of different reactions is assumed. By adjusting the ratio of protons to neutrons, which happens to be a free parameter in this calculation, the best possible agreement between theory and observations is obtained. This agreement is very good for the nuclei of the iron group; see Fig. 2.14.

It is worth noting here that one reaction which plays an important part in this calculation is the following:

$$e^-+e^+\rightarrow\nu+\bar{\nu},$$

that is, an electron (e^-) and a positron (e^+) annihilate each other and the resulting energy is used to produce the neutrino (ν)–antineutrino ($\bar{\nu}$) pairs. This is a weak interaction which could exist according to some theories but which has not been observed in the laboratory. The fact that its use in astrophysics seems to give the right answers (in terms of abundances) suggests that it must really take place,

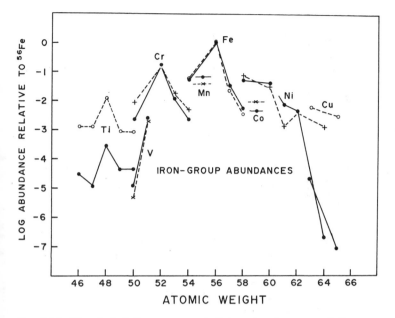

Fɪɢ. 2.14. The dotted lines represent Solar abundances and the continuous lines represent calculated abundances.

although the probability of its occurrence under laboratory conditions is small.

5. *The r,s,p-processes.* If the iron-group nuclei are the stablest, how can we form even heavier nuclei? Clearly the process cannot proceed simply along the lines of the α-process. The solution is provided via the mechanism of 'neutron capture'. Neutrons are chargeless particles and do not suffer from Coulomb repulsion. They can interact with heavy elements and help to produce heavier nuclei. This can happen in various ways.

The slow process (s-process) involves neutron capture at a rate *slower* than the β-decays which also take place in the nuclei. In β-decay the neutron is changed to a proton with the emission of an electron and an antineutrino. The s-process therefore results in nuclei of higher electric charge. In the rapid process (r-process) neutron capture is faster than β-decay. Thus, whereas the s-process produces proton-rich nuclei, the r-process produces neutron-rich nuclei.

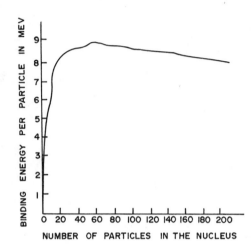

FIG. 2.15. Curve showing the binding energy per nuclear particle for various nuclei. The peak occurs in the region of the iron-group nuclei.

Apart from these two processes there is a rarer process which produces proton-rich isotopes by exposing the r-process and s-process material to a fast flux of protons or of high-energy photons. This is known as the p-process. The s-process takes place inside red giants, which have been formed from galactic material containing hydrogen, carbon, oxygen, neon, magnesium, and the intermediate iron-group elements. The r- and p-processes take place in the explosive envelopes of supernovae.

The supernovae serve the purpose of distributing the synthesized elements in the Galaxy. The 'second-generation' stars formed from the contaminated material already contain the heavy nuclei—unlike the 'first-generation' stars which were formed from pure hydrogen.

6. *The x-process.* Under this heading the Burbidges, Fowler, and Hoyle grouped together unknown processes which may be called upon to explain abundance anomalies, especially for deuterium (heavy hydrogen), lithium, beryllium, and boron. Processes beyond those operating in the stars might be required here.

To what extent stars are responsible for the elements we observe

can only be judged by comparing the detailed predictions of the
B²FH theory with the observed abundances of the various elements.
The astronomer has used various methods to estimate these abundances. The sources of information drawn upon include (a) the Sun
and other stars, (b) the gaseous nebulae, (c) the interstellar medium,
(d) the Earth, (e) the meteorites, and (f) cosmic rays. An overall
abundance curve has been constructed based on a detailed assess-
ment of the data. This curve is shown in Fig. 2.16. We notice that

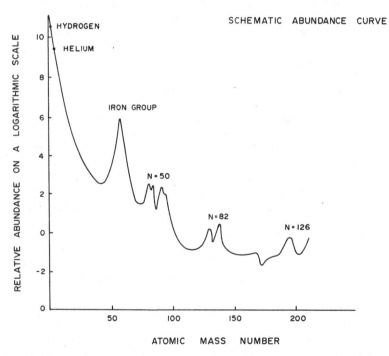

FIG. 2.16.

hydrogen and helium are by far the most abundant elements: the
rest may be no more than 2 per cent by mass. Although there is a
rapid decline in the abundance of elements heavier than helium, there
is a small peak near iron. Stellar nucleosynthesis is able to account

for these features successfully, except for the helium abundance (and we will return to this question in Chapter 7).

Planetary systems

Although astronomers have observed stars in plenty, as yet the direct evidence shows only one star with a planetary system, and that is the Sun. This might give the impression that the Sun is unique among all stars. While the present evidence does not rule out such a conclusion, the odds against its being true are very high.

First we must recognize that planets are not self-luminous. We are able to see them in our Solar System only because they reflect and scatter the light of the Sun. The brightness of a planet, therefore, is considerably less than that of the Sun. In the case of any other star, its planets, if they exist, will be so faint as to be invisible compared with the star. We must rely on indirect measurements to obtain evidence for their existence. Such evidence can come in the form of, say, a partial eclipse. If the orbit of the planet is such that it crosses our line of sight to the star, the planet will periodically obscure part of the star. Naturally a fairly good alignment of the orbit with respect to our line of sight is required, and this makes the arrangement somewhat improbable on a chance basis. That is, we have to be lucky to discover such a situation in any particular star. The other possibility lies in the disturbance of a star's own motion, caused by a fairly massive planet. A planet going round a star will exert a gravitational pull on it and disturb its straight-line motion. A periodic wobble in a star's motion could reveal the existence of a planet. So far both these methods have yielded somewhat tentative indications of planets associated with some stars.

Planets may be required as by-products of star formation. As we saw on p. 28, a cloud of dust and gas condensing into a star carries a lot of angular momentum, whereas the final fully formed star does not have much angular momentum. If the angular momentum was somehow conveyed from the star to a closely connected system, the mystery of the missing angular momentum would be solved. This other system could very well consist of planets orbiting round the star. Clearly such theories have first to be tried out in our own Solar System, and here, as yet, astronomers do not seem to have reached any clear-cut conclusions. However, the observations and results arising from recent space probes to the planets and flights to the Moon may bring about a serious revision of existing ideas.

Whatever mechanism is required for the formation of a planetary system, the number of stars in the Galaxy is so large that it is very improbable indeed that the Sun is unique in the Galaxy in having planets. On a crude estimate, the chance of this occurring is about one part in a hundred thousand million (10^{-11}), and this has to be reduced still further if we also demand that the Sun be unique in the Universe.

Is there extraterrestrial life?

This is a question which is often raised, and it is linked with the existence of other planetary systems. Life, of course, may exist in many forms, some unimaginable on the Earth. In our own system of planets there does not seem to be any advanced form of life anywhere except on the Earth. The most advanced creature on the Earth—the human being—can exist only under a fairly narrow set of environmental conditions, and unless these are reproduced on other planets —either in our planetary system or somewhere else—it is not possible to visualize a human existence elsewhere in the Universe.

But our biological knowledge is as yet too primitive for us to assert that life can exist only in the form seen on the Earth. Life may exist on other planets of other stars in radically different forms. It does not require the aid of a science-fiction writer to imagine superhuman existence elsewhere.

When the first pulsar was discovered the extreme regularity of its pulsed signals suggested the possibility that they were artificially generated by some super-civilization. However, such a civilization would have to exist on a planet orbiting round a star. And so the planet would have a motion relative to the Earth which would change magnitude and direction periodically. This would show up in the radiated signals through the Doppler effect (see p. 115), resulting in changes of frequency of the signals. No such effect was found, and so was discounted the possibility of a communication from 'little green men' in outer space! Nevertheless, interstellar communication on a grand scale cannot be ruled out, though at present we are probably technologically too primitive to intercept any such signals. After all, a primitive man in a jungle would be quite unaware of two explorers communicating with each other on a walkie-talkie. The ages of stars and planets could run into thousands of millions of years, and this is a long enough time for the evolution of life elsewhere to much more sophisticated forms than those seen on the Earth.

An epitaph for a star

Apart from spreading light in the Universe, the stars serve another useful purpose so far as we on the Earth are concerned. They provide us with the variety of elements which make life so interesting and liveable. Just imagine that the components of steel, which we find so useful today in a variety of ways, were cooked in some star at temperatures of several thousand million degrees and finally thrown out in a supernova explosion! But for the stars, we would not have our life-saving oxygen, or our valuable gold and diamonds. What is more, man himself is made from the materials synthesized in stars. It is fitting therefore that the Burbidges, Fowler, and Hoyle began their classic paper with the following quotation from Shakespeare:

> 'It is the stars,
> The stars above us, govern our conditions.'

3

Our Galaxy and Beyond

Although astronomical observations have often yielded unexpected results, sometimes astronomers have a premonition that their results may turn out to be important. Several precautions were taken when radio-astronomers Hazard, Mackey, and Shimmins carried out their lunar occultation observations in Australia, which were to lead to the discovery of the quasar 3C-273. Describing them in his book *Galaxies, Nuclei and Quasars*,† Sir Fred Hoyle writes: 'Several tons of metal were sawed off the telescope to permit observation at a lower angle of elevation than the normal operational range. For hours before the occultation all local radio stations broadcast repeated appeals: that no one should switch on a radio transmitter during the critical period of the observation. All roads leading anywhere near the telescope were patrolled to make sure that no cars were in motion in the vicinity. A final, somewhat macabre touch: after the observation Hazard and Bolton carried duplicate records back to Sydney, on separate planes.'

In the last chapter we looked at stars. We now turn to the bigger systems which appear to fill up the Universe as far as our instruments can probe. A survey of the Universe shows that stars are not distributed uniformly in space; they tend to occur in large groups, separated by almost empty regions. These groups are known as galaxies. The stars in our own neighbourhood, including the Sun, are members of a galaxy which we shall refer to as our own Galaxy or the Galaxy popularly identified as the Milky Way. Our Galaxy is only one of the various types of galaxies found in the Universe. Apart from galaxies the Universe contains other more spectacular objects, such as 'quasi-stellar objects' (or quasars). In this chapter I propose to describe some interesting features of these objects, most of which are so remote and faint that the human eye is incapable of seeing them. Of course, of all

† See p. 50, 1965 Edition, Heinemann, London.

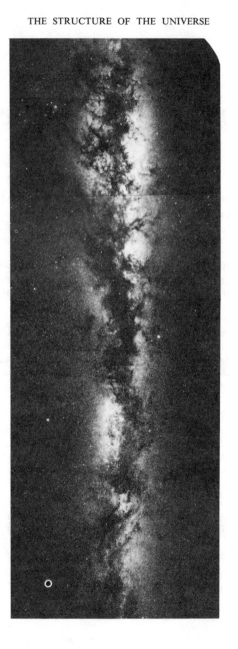

FIG. 3.1.   The Milky Way. (Photograph from the Hale Observatories.)

the various systems in the Universe the one about which we have most detailed information is our own Galaxy, and it is best to begin with a description of what we know about it.

Our Galaxy

What does our Galaxy look like? Unlike other galaxies, which we can look at from the outside, we have to build up a picture of our Galaxy from observations taken from the inside. This is more difficult, which partly explains why, until the beginning of the present century, most astronomers believed that our Galaxy made up the entire Universe. Only with careful observations was it possible to make out that many of the objects in the sky did not form part of our Galaxy but lay well beyond. We shall return to this point again later.

Fig. 3.1 shows a composite photograph of the Milky Way system. The photograph is made up by joining together photographs taken in many different directions. The Milky Way is a band of stars going right round us. In fact it represents the projection of the so-called 'galactic disc'. The overall shape of the Galaxy is like that of a bun (see Fig. 3.2), consisting of a central disc of about 30 kpc (kpc =

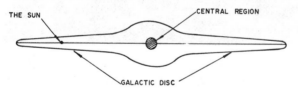

FIG. 3.2. Schematic diagram of the Galaxy.

kiloparsec) diameter, representing a relatively dense concentration of stars, together with a bubble (or halo) representing a less-populated region. We ourselves are located in the disc, some 10 kpc away from the Galactic centre. To get some idea of what our Galaxy looks like from outside, we need only look at our neighbouring galaxy M31 in Andromeda (M31, or Messier 31, is a catalogue number for the galaxy), a picture of which is shown in Fig. 3.3.

Although our Galaxy contains some hundred thousand million (10^{11}) stars, it also has two other important constituents—dust and gas. These make up the 'interstellar medium', which is of great interest to astrophysicists, even though it makes up only about 2 per

FIG. 3.3.   The Andromeda Nebula. (Photograph from the Hale Observatories. (Copyright by California Institute of Technology
and Carnegie Institute of Washington.))

cent of the mass of the entire Galaxy. Apart from the interstellar
medium, the Galaxy contains a magnetic field, which also has a
significant role to play inside the Galaxy.

Stars

Surveys of star densities in different directions show that stars tend
to be concentrated in the plane of the Milky Way. Also, they extend
to greater distances along the plane than in directions perpendicular
to it. This is similar to star distributions in several other nearby
galaxies. Fig. 3.4 is a graph showing how the relative star density
falls off as we move away from the galactic plane. The curve represents the overall average for all stars. There are in fact differences in

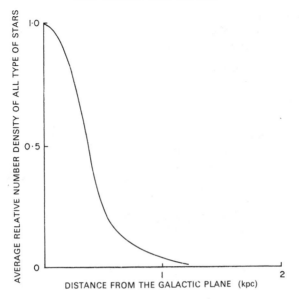

FIG. 3.4.   Star density perpendicular to the galactic plane.

the distribution of stars of different magnitudes (see Chapter 2 for a
discussion of stellar magnitudes), but the general pattern is the same.
There is a significant drop in the star density as we go more than
1 kpc away from the galactic plane.

Copernicus first suggested that the Sun is at rest while the Earth
and the other planets move round it (before Copernicus it was commonly believed that the Earth stood still). Present observations show
that, viewed from the galactic standpoint, even the Sun is not at rest
but is slowly moving round the Galaxy. By examining the motions
of nearby stars relative to the Sun, William Herschel concluded, as
far back as 1787, that the Sun was slowly moving in the direction of
the Hercules constellation. Modern measurements differ from this
conclusion only within a few degrees (see Box 3.1).

Box 3.1 Sun's motion in the Galaxy

The motion of a star relative to the Sun can be thought of as having
two components: one, the radial component, denotes its velocity
away from the Sun while the other component denotes the star's

velocity in a direction perpendicular to the line joining the Sun to the star. The latter is called the 'proper motion' and can be measured after a painstaking comparison of the star's position in the sky at intervals of several years or decades.

Stars in the neighbourhood of the Sun show proper motions which suggests that the Sun itself may have some motion through space. Motion relative to what? Clearly, a standard of rest, or a reference frame, has to be defined, and this is provided by the background of the stars on the assumption that they are moving in random directions. If the Sun were truly at rest relative to this background, the average motion of stars in its neighbourhood would be zero as required by the randomness of stellar motions. If not, there should emerge an average background velocity equal and opposite to the velocity of the Sun relative to the background. In other words, if the Sun's velocity is directed towards a point called the 'apex' in space, the stars should appear to move away from the apex and towards the antapex in the opposite direction (see Fig. 3.5.)

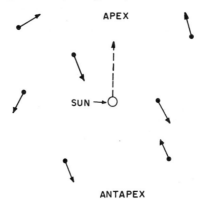

FIG. 3.5.   The average velocity of the stars in the solar neighbourhood will be directed towards the antapex.

By 1787, from his studies of the proper motions of some 36 stars, William Herschel determined the apex of the Sun's motion in the direction of the Hercules constellation. The magnitude of the velocity was determined later by Campbell of the Lick Observatory, after an analysis of the data on 2149 stars. He obtained a value of 19.6 km s^{-1}. The value, of course, does not reflect the streaming motion of the Sun round the Galaxy. This was determined by other methods, as discussed in Box 3.2.

After many decades of work involving measurements of star motions in the Galaxy, a coherent picture has finally emerged. In 1926 Lindblad, a Swedish astronomer, suggested that the Milky Way as a whole is rotating. A few years later the Dutch astronomer Jan Oort succeeded in giving a simple calculation of the speed of rotation of the Galaxy, as well as its mass (see Box 3.2 for further details).

---

†Box 3.2 Rotation and mass of the Galaxy

Stars in the Galaxy exhibit a systematic streaming motion, over and above small random velocities in different directions. The orbit of a typical star in this streaming motion is elliptical, with properties similar to the elliptical orbits of planets. Most orbits like that of the Sun, are nearly circular, while some are highly eccentric. These latter orbits (like those of comets) are of high-velocity stars. All the orbits are about the centre of the Galaxy, which lies in the constellation of Sagittarius.

Suppose all orbits are circular and in the plane of the Galaxy. Consider the orbits of stars in the neighbourhood of the Sun. Assuming that each of these orbits is determined by the gravitational pull of the mass of the Galaxy which falls inside the orbit, the controlling mass for all these orbits is nearly the same and may be regarded as concentrated at the centre of the Galaxy. This assumption is justified in view of the fact that a great deal of mass is concentrated at the Galactic centre. If M is the mass involved, R the radius of the orbit, and V the star's velocity, we get

$$V^2 = \frac{GM}{R}. \tag{1}$$

(1) is obtained by equating the centrifugal force in the circular motion with the gravitational pull of mass M.

From this it is clear that as R increases V decreases. Also, as shown in Fig. 3.6, two sets of stars will show zero radial motion with regard to the Sun. The first set lies on the radial direction CS from the galactic centre to the Sun. The second lies on the tangential direction Y'SY, but at distances from the Sun small compared to the radius R of the Sun's orbit. Stars in intermediate directions will show non-zero radial motion. Also, as we go away from the Sun, the relative velocity of the star with respect to the Sun increases. For small distances the effect is linear. Oort expressed the transverse velocity of the star in the following form

$$T = r\{B + A\cos 2(l - l_0)\}. \tag{2}$$

Here r is the distance of the star from the Sun, l its galactic longitude, and l_0 is the galactic longitude of the galactic centre. A and B are constants called Oort's constants. Comparison with observed proper motions showed a good agreement with the above formula. A and B are determined by observations and are given by

$A = (18 \cdot 2 \pm 0 \cdot 9)$ km s^{-1} kpc^{-1}, $B = (-8 \pm 2)$ km s^{-1} kpc^{-1}.

The motion is clockwise as seen from the north pole of the Galaxy.

Using these results it is possible to work out the Sun's velocity relative to the galactic centre. The orbital velocity of the Sun is estimated at 250 km s^{-1}. Also the distance of the Sun from the centre of the Galaxy has been measured, and is about 10 kpc. Then from (1) we get the mass M as

$$M = \frac{V^2 R}{G} \simeq 1 \cdot 4 \times 10^{11} \, M_\odot. \qquad (3)$$

From this calculation the orbital velocity of the Sun comes out as about 250 km s^{-1}. At this velocity the Sun (and the associated Solar System) takes about 200 million years to go round the Galaxy.

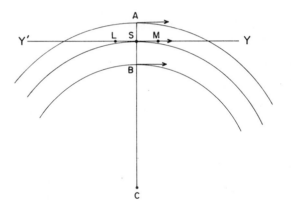

Fig. 3.6.

Dust

When we look at photographs of the Milky Way system or of other galaxies one striking feature is the mixture of bright and dark regions (see the photographs in Fig. 3.7). While the bright regions may be identified with collections of stars, what interpretation can be given to the dark regions? The early astronomers believed the dark regions to be genuinely empty, and thus maintained that interstellar space is

FIG. 3.7. (a) The Horsehead Nebula in Orion. (Photograph from the Hale Observatories.) (b) Band of dust across the galaxy NGC 4594 in Virgo. (Photograph from the Hale Observatories.)

nothing but a void. However, photographic evidence gradually dis-
pelled this belief and pointed to the existence of dark obscuring
clouds in the Galaxy. These clouds vary in their obscuring power—
some are totally dark, while some permit a certain amount of light
to pass through them.

What are these clouds made of? This question has been answered
by calculating the degree of absorption produced by different materials of various sizes. The size of the particles forming the clouds
plays an important part. In Fig. 3.8(a) we have a sphere of a certain
radius R containing an absorbing matter, while in Fig. 3.8(b) the
same matter is divided into eight smaller spheres, each of radius $\frac{1}{2}R$.

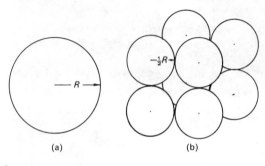

FIG. 3.8.

The combined front, or 'cross-section', presented by the eight small
spheres is twice as much as the cross-section of the single larger
sphere. The absorbing power of a system increases with its cross-
section. Therefore the smaller spheres taken together act as better
absorbers than the larger sphere. So obscuring clouds act more
efficiently if they are made up of small particles rather than huge
chunks of matter. However, extremely small particles also are not
good absorbers, and thus there is an optimum size for absorption.
The absorption is most efficient when the size of the absorbing particle is comparable to the wavelength of the light to be absorbed. For
visible light the best particle size is therefore in the region of a micro-
metre or a micron ($= 1\ \mu m = 10^{-6}$ m), and the obscuring material is
called 'dust'. As we decrease the particles' sizes further, they do not obscure so efficiently. Thus when we get down to gases composed of

atoms or molecules, they act as poor obscuring agents. Such gases do produce absorption, but of a different sort, and this will be discussed later.

When interstellar dust is in its diffuse form, it affects the apparent brightness of distant stars. A star appears less bright than expected because of diffuse interstellar absorption. Consequently, if we ignore the existence of the dust, we are likely to conclude that the star is further away than it actually is. Before the existence of interstellar dust was recognized, astronomers grossly over-estimated stellar distances.

Apart from causing obscuration, the dust also produces a change of colour. This phenomenon is known as 'Rayleigh scattering', and takes place when light impinges on a dust particle. The light gets scattered, and in the process loses a good deal of its shorter-wavelength components. This has the effect of increasing its overall wavelength, that is, making it redder in colour. This result is known as 'interstellar reddening'.

Absorption and scattering by the dust particles provides us with important information about their size and number density. These, in turn, help the astronomer to decide the composition of the dust grains. At present a great deal of work is being done on trying to determine the nature of interstellar dust. The composition of the dust must be related to the abundances of elements (see p. 48). The preponderance of hydrogen, oxygen, carbon, and nitrogen suggests that the dust must be made of a condensed form of these gases. The early work in this field was centred round the hypothesis that the dust grains are made of ice, but more recent work shows that the composition of a grain may be more complicated. Among various models considered there are grains with graphite (a crystalline form of carbon) cores and ice mantles, grains of solid hydrogen, and silicon grains.

Gas.

While dust grains produce absorption in the continuum emission of a star, gases produce absorption lines (see Box 2.4, p. 23) in the spectra of stars. The first indications of this appeared at the beginning of this century, when absorption lines of ionized calcium—the so-called H and K lines of CaII—were observed in the spectrum of the double star δ-Orionis. Since then stellar spectra have revealed the existence of several elements and radicals (chemical combinations of two or more elements) in interstellar space (see Fig. 3.9). These include NaI, CaI, CaII, KI, CH, and CN. Radio-astronomical techniques have also led to the discovery of H, OH, D, NH_3, HCHO, HCN, CH_3CN, etc.

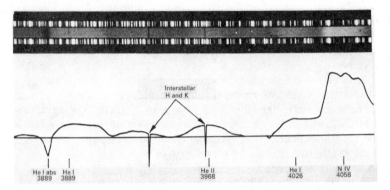

F IG. 3.9.   Example of interstellar lines of ionized calcium (in the spectrum of HD 192163). The curve below the spectrum shows the change in the intensity of light received, with the sharp clips representing the absorption of H and K lines. (Photograph from Kodaikanal Observatory, Indian Institute of Astrophysics.)

Gas appears in a more spectacular form than dust, in the so-called 'emission nebulae'. These nebulae are enormous masses of gas which absorb ultraviolet radiation from stars and radiate it again as visible light. Two such nebulae are shown in Fig. 3.10. These bright nebulae are usually formed around stars, and are mainly composed of hydrogen.

Hydrogen occurs in two forms in the Galaxy. The so-called 'HI regions' contain neutral hydrogen while the 'HII regions' contain ionized hydrogen. The latter type is a zone formed around a star where the ultraviolet light from the star heats up and ionizes the surrounding hydrogen. An HI region, on the other hand, is cool, and it is in these regions that new stars are formed (see p. 26).

Radio-astronomy has played an important part in galactic explorations. For example, 21-cm radio wavelength observations are useful for detecting neutral hydrogen. This radiation arises when the

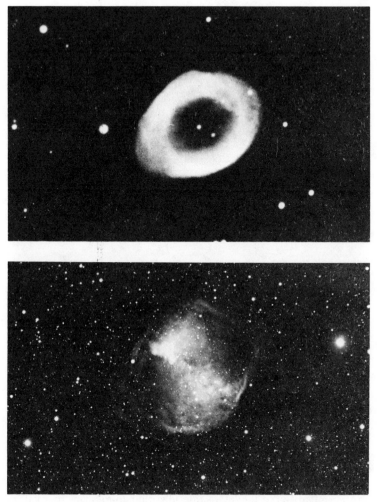

FIG. 3.10. (a) The Ring Nebula. (Photograph from the Hale Observatories. (Copyright by California Institute of Technology and Carnegie Institute of Washington.)) (b) The Dumbbell Nebula. (Photograph from the Hale Observatories. (Copyright by California Institute of Technology and Carnegie Institute of Washington.))

orbiting electron in a neutral hydrogen atom reverses its spin, from a direction parallel to the spin of the proton at the nucleus to the opposite direction. The electron loses energy while changing (in relation to the nucleus) its spinning direction, and in this process emits radiation of wavelength approximately 21 cm.

External galaxies

The vast multitude of other galaxies found in the Universe have been classified into various categories, according to their shapes and other characteristics. Since some of the characteristics may be shared by galaxies of different classes, these classes are not necessarily mutually exclusive.

Spiral galaxies

Our Galaxy belongs to the category known as 'spiral galaxies'. However, it is obviously difficult to make out what our Galaxy looks like by observations taken from inside. Indeed it took many years of hard work by astronomers to establish the spiral structure of our Galaxy. But if we look at other galaxies it is easy to make out this spiral shape (see Fig. 3.11).

The spirals are classified in a sequence Sa, Sb, Sc, corresponding to the relative importance of the central bulge and the disc (which has the spiral structure). Thus the bulge is largest in Sa and weakest in Sc. Our own Galaxy is close to the class Sb. Apart from this there are lenticular galaxies S0, which resemble the spirals in many respects but which are very much dominated by the bulge. Also there are barred spirals (see Fig. 3.12), which are dominated by a bar-like structure.

How did the spiral structure arise? As yet we have no satisfactory explanation of this phenomenon. In the mid-1950s, Chandrasekhar and Fermi advanced the interesting idea that spiral arms in a galaxy like ours arise from the magnetic field. If the magnetic field lines are spiral in structure (Fig. 3.13) matter may tend to follow the lines. While this may be plausible for electrically charged particles, it is difficult to see why stars as a whole should be guided by magnetic lines of force. Nevertheless, whatever the ultimate success or failure of this theory, it is a bold attempt to understand galactic structure.

FIG. 3.11. Spiral galaxies. (Photograph from the Hale Observatories.)

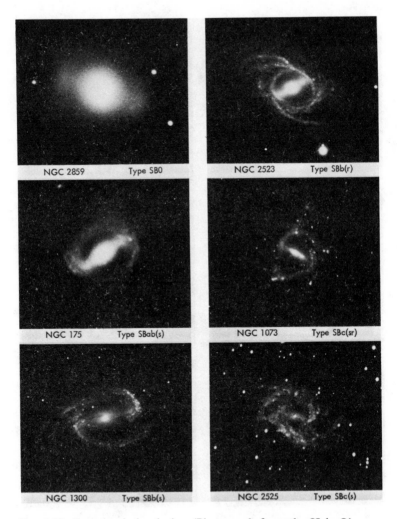

FIG. 3.12. Barred spiral galaxies. (Photograph from the Hale Observatories.)

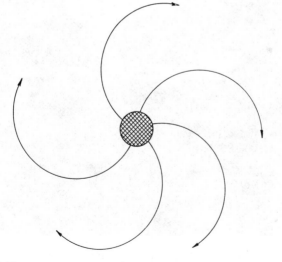

FIG. 3.13.

FIG. 3.14. The Elliptical (E0) galaxy NGC 4486. (Photograph from the Hale Observatories.)

Elliptical galaxies

As the name suggests these galaxies appear to us to be elliptical in shape. Like the spirals, the ellipticals are classified as E0, E1, . . ., E7, depending on their eccentricity (see Fig. 3.14). Thus E0 is nearly round. It is also the most massive. A typical E0 can be as much as 10 times more massive than our Galaxy. A typical cluster of galaxies is usually made up of a number of spirals and perhaps a few ellipticals, dominated by an E0-type galaxy. And such a massive E0 is nearly uniform in its mass and light emission, in all such clusters. This circumstance has been used by cosmologists in testing the law of expansion of the Universe (see p. 210 for details).

Apart from their masses the ellipticals differ from the spirals in other respects. The mass-to-light-emission ratio for ellipticals is significantly higher than that for the spirals. This means that for the same quantity of matter, the spirals manage to emit much more light

POLARIZED LIGHT

EV 200°   EV 290°

EV 245°   EV 335°

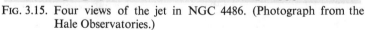

FIG. 3.15. Four views of the jet in NGC 4486. (Photograph from the Hale Observatories.)

than the ellipticals. This is largely because ellipticals contain mostly old stars, stars which are approaching the end of their evolution (see Chapter 2) and hence are not as bright as when they were young. The spirals mainly contain stars in earlier phases of evolution, and thus they tend to emit more light. There is also very little gas or dust in the ellipticals, in contrast to the amount in spirals.

This might give the impression that ellipticals are relatively quiet systems, but evidence to the contrary has been accumulating over the last few years. The central or nuclear regions of the ellipticals appear to be seats of violent activity. In Fig. 3.15 is shown the same E0 galaxy NGC 4486† which was shown in Fig. 3.14. The photographs show a jet coming out of the galaxy, indicating some violent activity at the centre. This galaxy is also known to emit radio noise.

Irregular galaxies

These do not fall under either of the two schemes described above. Their shapes and composition vary enormously, so that they cannot be placed under any single category. A comprehensive atlas of 'peculiar' galaxies has been prepared by Vorontsov Velyaminov. In Fig. 3.16, an example is given of such galaxies. Various conjectures have been made about the nature of these systems, but as yet they have not found a proper place in the general extragalactic scheme.

The red-shift

Before proceeding further with the list of galaxy classification it is necessary to understand an important property shared by all three of the classes of galaxies described above. This is known as the 'red-shift' and is the systematic increase in the wavelength of light coming from a typical galaxy. Why the name red-shift, and how is it manifested in the astronomical data?

When the astronomer examines the spectrum of light coming from an extragalactic object, he usually finds emission or absorption lines (or both). However, these lines do not have the wavelengths expected —rather, the wavelengths are increased in a fixed ratio. So, if the expected wavelength is λ_0, the observed wavelength appears to be

$$\lambda = \lambda_0(1+z),$$

where z is positive. Of course, in arriving at this conclusion the astronomer must have a fairly clear idea of what λ_0 should be .In

† The letters NGC stand for the New General Catalogue.

FIG. 3.16. M82: an irregular galaxy. (Photograph from the Hale
Observatories. (Copyright by California Institute of Technology and Carnegie Institute of Washington.))

this he is guided by the spectroscopic information from similar
nearby objects. For example, the presence of hydrogen is reflected in
the $H\alpha$, $H\beta$, $H\gamma$ lines of known wavelengths. Suppose three lines are
found with the same wavelength ratios as of the above three lines but
not with the same wavelengths. Then the astronomer concludes that
the wavelengths are all changed by the same factor $(1+z)$. The
wavelength ratios therefore provide the main clue.

The name red-shift arises because in the visible part of the spectrum
red is the colour with the longest wavelength (see Box 2.2, p. 18).
A positive z factor applied to light of any other colour would increase
its wavelength, and thus shift it towards the red. The increase in
wavelength results therefore in the shifting of the original spectrum
towards the red end. If z were negative the spectrum would have

shifted towards the opposite end, that is, towards the colours blue and violet. Hence negative z is interpreted as 'blue-shift' or 'violet-shift'.

In later chapters I will explain the various possible interpretations of the red-shifts, but now we must return to the types of galaxies found in the Universe.

Radio galaxies

The classification of galaxies described so far is based solely on their characteristics as seen by visible light. Until about the time of the Second World War optical astronomy was the only source of information about the Universe. This monopoly has since been gradually eroded. Radio-astronomy was the first gate-crasher, and it arrived in an impressive way. In the 1940s and the early 1950s astronomers did not expect any very dramatic information to come from radio observations. But in this they were proved wrong.

In 1946, Hey, Parsons, and Phillips detected strong radio emission from a small area in the constellation of Cygnus. This was followed by other observations of radio emission. The sources seemed to be of two classes. Those in the first class seemed to be concentrated in the plane of the Galaxy, while those of the second class were distributed isotropically, that is, equally in all directions. The former sources could therefore be associated with the Galaxy itself—but what about the latter?

Most astronomers believed that they were radio stars distributed over a fairly nearby region in the Galaxy. This view was challenged by Thomas Gold, who argued that the distribution of such stars implied that they were distant extragalactic sources. Eventually Gold was proved right. Some of the sources were identified optically with external galaxies. This process of optical identification is an example of the co-operation between optical astronomers and radio-astronomers. A radio-astronomer measures the position of the source in the sky and communicates this information to his optical counterpart. The position is quoted not as a precise point but as a small rectangle within which the source is expected to lie. The optical astronomer then examines his plates to see if a possible object is visible in that rectangle. If the rectangle is very small (that is, if the radio position has been measured very accurately) the identification with an optical object is easier and less ambiguous.

This process was adopted for the source detected in Cygnus.

Graham Smith, working with the radio telescope at Cambridge, U.K., measured the position of the source accurately and communicated it to Walter Baade at the Mt. Wilson and Palomar Observatories (now

FIG. 3.17. The Cygnus. A source of radio noise. (Photograph from the Hale Observatories.)

the Hale Observatories) in California, U.S.A. Baade identified the source, now known as Cygnus A, with what looked like a pair of colliding galaxies about 500 million light-years away (see Fig. 3.17). So Cygnus A, along with most of the other sources of the second class, turned out to be an extragalactic object.

This discovery of Cygnus A led to the hypothesis that radio sources arise in extragalactic space when two galaxies collide. The collision hypothesis looked attractive in the sense that it called upon a dramatic explanation for a dramatic event. Subsequent observations were to show that this hypothesis is not a likely explanation.

Fig. 3.18. Galaxy NGC 5128 identified with the radio source Centaurus A. (Photograph from the Hale Observatories.)

This conclusion is arrived at after the optical identification of several radio sources. In some cases the optical objects were double elliptical galaxies well apart from each other; in other cases there was a single optical object. In the early 1960s, Geoffrey and Margaret Burbridge examined the gas motions in the object NGC 5128 (see Fig. 3.18), which is identified with the radio source Centaurus A. They found the motions too small to represent collisions. If we look at the photograph of this object we find a band of dust separating two bright portions. If the object were very distant, like Cygnus A, we might not see the band clearly, and would imagine the object as made up of two colliding galaxies. Thus even Cygnus A is more likely to derive its energy from some other source than the collision of galaxies.

The collision hypothesis also proved troublesome when attempts were made to explain the enormous energy released in the radio sources. The power radiated by the source Cygnus A is estimated to be about a million million million million million megawatts (10^{36} W). It is difficult to get so much power out of collision, even on a galactic scale.

According to the present picture a radio source is typically a double source, with its two components separated by a distance ranging from 35 kpc (for a small system like 3C-31†) to as much as 750 kpc (for Centaurus A). There is usually a single optical object—a galaxy—located between the two components (see Fig. 3.19). This

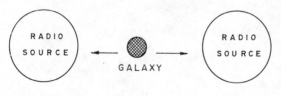

FIG. 3.19.

may be a somewhat over-simplified picture, but it serves as a starting point for most models of radio sources. We shall return to these models later.

† 3C-31 means the 31st source in the 3rd Cambridge Catalogue of radio sources.

FIG. 3.20 (a). Seyfert Galaxy NGC 4151. (Photograph from the Hale Observatories.)

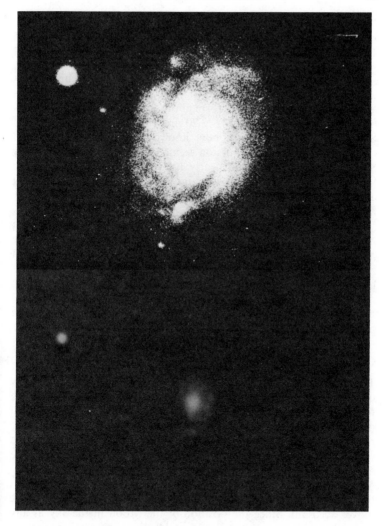

FIG. 3.20 (b). Seyfert Galaxy NGC 1068. The lower half of the picture is
        underexposed to show the central nucleus. (Photograph
        from Kodaikanal Observatory, Indian Institute of Astro-
        physics.)

Seyfert and N-galaxies

In 1943 Seyfert investigated a special class of galaxies which is characterized by the fact that the nuclei of such galaxies seem to be very much like stars. The light from the nuclei of the Seyfert galaxies —as they are now called—shows broad emission lines. This characteristic sets the Seyferts apart from the rest of the galaxies. It is estimated that these galaxies form about 1–2 per cent of the population of spiral galaxies. In Fig. 3.20 are shown two photographs of Seyfert galaxies which are now known to emit strong infrared radiation. This infrared emission seems to be associated with many Seyfert galaxies. Some emit nearly 100 times more radiation in the infrared than our Galaxy does in the visual range. The recent developments in infrared astronomy have therefore added a new facility to galactic studies.

N-galaxies are very similar to Seyferts. They also have brilliant nuclei which show up against a considerably fainter background. The main difference is that most N-type galaxies are strong emitters of radio waves. Also, they have large red-shifts, whereas the Seyferts have small red-shifts. Indeed the nuclei of N-type galaxies are in many respects similar to a more remarkable class of astronomical objects which were discovered in early 1963; these objects became popularly known as 'quasars'.

Quasi-stellar objects (QSOs)

The discovery of the first quasi-stellar objects came in a very unexpected fashion. In 1962, Cyril Hazard, working with M. B. Mackay and A. J. Shimmins at the Parkes radio astronomical observatory in Australia, accurately measured the position of the radio source 3C-273. The method used in this measurement was a relatively new one at that time, although nowadays it has become quite common. It involves lunar occultation, that is, the transit of the Moon across the line of sight of the radio source. The radio output from the source drops sharply (see Fig. 3.21) when the lunar disc moves across it. Since the position of the Moon in the sky at any given time is known very accurately, this observation enables astronomers to pinpoint the position of the source in the sky to within very small errors. 3C-273 is a double source, and the positions of both of its components were determined in this way. This information was communicated to optical astronomers at the Hale Observa-

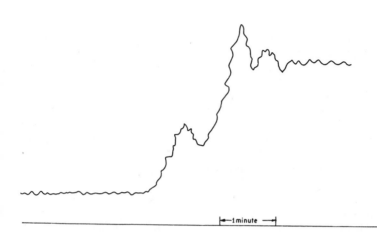

FIG. 3.21. Schematic occultation curve showing the change in the output of the radio source 3C–273. The two peaks correspond
to the two components of the source. (After Hazard, Mackey,
and Shimmins observations.)

tories in Southern California. A careful examination of the photographic plate showed that a star-like object of 13th magnitude was
located on one of the components of the radio source. In Fig. 3.22

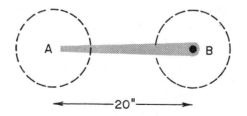

FIG. 3.22. Schematic diagram of the QSO 3C–273.

we have a schematic diagram of the source 3C-273 and the star-like
object identified with it. The star-like object has a jet which reaches to

within about 20″ away from it.* The other end of the jet coincides with the second component of the double radio source 3C-273.

This optical object was, of course, known previously; and it was assumed to be a star in our Galaxy. However, its optical identification with 3C-273 led to its further investigations. When its spectrum was examined it disclosed four emission lines—but their wavelengths did not correspond to those of any known spectral lines. This puzzle was resolved when it was assumed that the spectrum may be red-shifted. Maarten Schmidt of California Institute of Technology confirmed this by measuring the red-shift (z) as 0·158. That is, the wavelength of all four lines (OIII, Hβ, Hγ, and Hδ) had been increased by 15·8 per cent. A further check on this was provided by Oke, also of the California Institute of Technology, by measuring the wavelength of the Hα line by spectrum-scanning methods. This wavelength had also increased by the same proportion. A similar analysis of another star-like object, identified around the same time with 3C-48, revealed a red-shift of $z = 0·367$.

This discovery caused great excitement, for normal stars in our Galaxy do not have such high red-shifts. If red-shifts (as discussed in Chapter 4) imply motion away from us, the 'star' identified with 3C-273 must be moving away with nearly one-sixth of the velocity of light. This is far in excess of the typical stellar velocities in our Galaxy. If, on the other hand, the red-shift arises from the expansion of the Universe (see p. 121) the object must be extragalactic and situated a long way away. This meant that the object was intrinsically very bright—yet at the same time highly compact, as indicated by its star-like image. Calculations showed that to produce so much radiation the object must contain at least a million solar masses in a comparatively small region of a few parsecs. A third possible explanation of the red-shift (see p. 157) required the object to possess a very strong gravitational field—a field far stronger than had been detected anywhere else at that time. In any case, the astronomers realized, they had discovered a new and unusual type of object in the Universe. Because such an object looked like a star, but was also a strong radio source, it was named a 'quasi-stellar radio source'. For popular literature the name was shortened to 'quasar'†.

* By 20″ is meant 20 seconds of arc. A second of arc is a measure of the angle subtended by the source at the eye and is $\frac{1}{60}$ of a minute of arc and $\frac{1}{3600}$ of a degree.
† Now this term is also accepted in scientific literature.

Although the discovery of the first two quasars was accidental, a study of their properties gave astronomers some guidance about what to look for in their search for further quasars. By 1967 about 150 quasars had been discovered, and their general characteristics were well documented. By 1971 this number had risen close to 200 and is close to 500 now. The general properties of quasars can be summarized as follows:

1. They are star-like objects *often* identified with radio sources. The word 'often' emphasizes the fact that, although the early quasars were also radio sources, many were subsequently discovered without radio emission. Hence the name 'quasi-stellar radio source' was subsequently changed to 'quasi-stellar object', or just QSO.

2. Many of them have a variable output of optical and of radio emission (if they are radio sources). These variations are over a very short time-scale—of months or even days (see Fig. 3.23). This is quite a different behaviour from that of a galaxy or a radio galaxy, whose output of radiation does not change over such a short time scale.

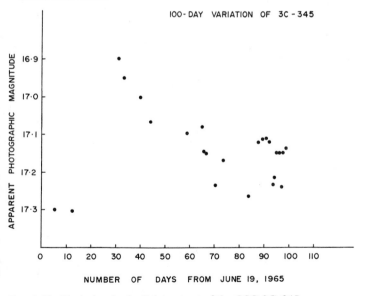

FIG. 3.23. Variation in the light output of the QSO 3C–345.

3. They have a large ultraviolet flux of radiation. This was one of the principal observed characteristics of early QSOs which helped in the discovery of further QSOs (see Box 3.3).

4. The spectrum of a QSO has broad emission lines. In many QSOs, absorption lines are also present. These play an important part in the construction of models for these objects.

5. The spectral lines show large red-shifts. The largest red-shift known for a galaxy is $z = 0.63$ (this is for the galaxy identified with the radio source 3C-123), but for QSOs red-shifts close to 2 are quite common, and at the time of writing three QSOs with red-shifts in excess of 3 are known. These large red-shifts are perhaps the most puzzling aspect of QSOs, and will be discussed at length in Chapter 7.

Box 3.3 The ultraviolet excess in QSOs

Although QSOs are star-like in appearance, there is one property which has been useful in distinguishing them from stars in the Galaxy. This is illustrated in Fig. 3.24, which is a two-colour plot for QSOs.

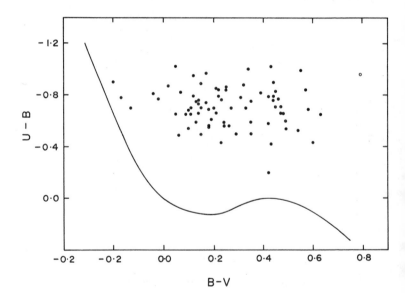

FIG. 3.24.

On the horizontal axis is B–V, that is, the colour index showing the difference between the blue and the visual magnitudes (see Chapter 2 for definitions of magnitudes, colours, etc.) On the vertical axis is U–B, that is, the colour index showing the magnitude difference between ultraviolet and blue. The continuous line is the characteristic line for galactic stars, while the QSOs are dotted well above it. This shows that, for the same B–V, the QSOs have significantly more U–B than for the galactic stars. In other words, the QSOs seem to emit comparatively much more in ultraviolet than do typical stars. Hence, in looking for new QSOs, the astronomer picks upon star-like objects showing a marked ultraviolet radiation excess.

Because of their star-like appearance QSOs are somewhat similar to N-type galaxies. Some astronomers have suggested an evolutionary link between the two. The main difference between the two is the absence of a fainter background in the case of a QSO. The red-shifts of QSOs are also generally higher than those of N-type galaxies.

What keeps a QSO shining? Because of its star-like appearance it might be thought that nuclear reactions are mainly responsible, This, however, is not the case, and it is believed that gravitation plays a dominant role in determining the behaviour of a QSO; but the exact nature of this mechanism is not yet known.

High-energy astrophysics

The observations of extragalactic astronomy, which have been briefly described above, have shown that enormous reservoirs of matter and energy exist in the space beyond our Galaxy. These reservoirs are of either a quiet or an explosive nature. Our own Galaxy, although it contains matter in excess of a hundred thousand million Suns, is a relatively quiet object. On the other hand, Centaurus A (Fig. 3.18) clearly shows signs of a large-scale explosion at its centre. It is exploding objects like Centaurus A that have presented many puzzles to astronomers. How did such explosive situations arise in the Universe? What is the source of energy behind them?

Associated with these questions is the problem of high-energy particles and light radiation. Cosmic rays contain particles of extremely high energy. The highest-energy particles have energy as high as a thousand million million million electronvolts (10^{15} MeV).† The

† This may be compared with the energy of about 1000 MeV which is stored in a hydrgoen atom at rest.

man-made accelerators on the Earth reach barely a thousand-millionth part of this figure. Then we have X-rays and γ-rays, which are electromagnetic radiations at a very high energy (see Box 3.4). What is the nature of their origin?

Box 3.4 The electromagnetic radiation at different frequencies

We have already seen (in Box 2.2) that electromagnetic radiation in the visual range (that is, in the wavelength range 4000–8000 Å) manifests itself as light of different colours. Beyond this range it is not visible, but it is nevertheless very important for the study of the Universe. The following table gives the names given to this radiation in different ranges of wavelength.

Name of radiation	Wavelength range	Frequency range
Radio	10^{-1}m–infinity	$(0-3 \times 10^9)$ Hz
Microwave and millimetre-wave	3×10^{-4}m–10^{-1}m	$(3 \times 10^9-10^{12})$ Hz
Infrared	8×10^{-7}m–3×10^{-4}m	$(10^{12}-3 \cdot 7 \times 10^{14})$ Hz
Visible light	4×10^{-7}m–8×10^{-7}m	$(3 \cdot 7 \times 10^{14}-7 \cdot 5 \times 10^{14})$ Hz
Ultraviolet	3×10^{-10}m–4×10^{-7}m	$(7 \cdot 5 \times 10^{14}-10^{18})$ Hz
X-rays	3×10^{-12}m–3×10^{-10}m	$(10^{18}-10^{20})$ Hz (4 keV–4 MeV)
γ-rays	$0-3 \times 10^{-12}$m	10^{20}–infinity (0·4 MeV upwards)

These ranges are not clear-cut—there is an overlap at the borders.

In some cases, the energy is given in brackets against frequency. This is according to Planck's formula relating energy E to frequency ν:

$$E = h\nu,$$

where h is Planck's constant. The 'electronvolt', that is, the work done in moving an electron against a potential barrier of one volt, is a suitable unit in this case.

Except for light, the radio frequencies, and parts of the infrared range, most other ranges are inaccessible to ground-based astronomy. This is because the atmosphere absorbs radiation in these other ranges. To make use of these ranges, the astronomer has to use balloons, rockets, or satellite-based instruments. And this has now become possible thanks to recent technological advances.

All these questions now form part of the subject of 'high-energy astrophysics'. The words 'high energy' can be used in one of two senses. They could stand for enormous reservoirs of energy or they could stand for particles and radiation of high energy. In the latter sense the subject is an extension of the term 'high-energy physics' which is used by the laboratory scientist. The high-energy physicist studies the behaviour of matter and radiation at high energies. But the energy range encountered by the astronomer is far in excess of the energy range available to the high-energy physicist. Before we can understand the work of the high-energy astrophysicist, we must understand something of the types of radiation which have found wide applications in high-energy astrophysics. These are quite different from the thermal radiation mentioned when we were discussing the composition of stars (see Chapter 2).

Synchrotron emission

To understand the basic properties of this type of radiation we have to consider what happens when an electric charge moves. If the charge is at rest or if it is moving in a straight line with uniform speed, it generates an electromagnetic disturbance which is unchanging with time. Such a disturbance falls off rapidly (as the inverse square of the distance) away from the electric charge, so that at large distance from it there is very little effect felt. If, however, the charge is made to accelerate, either by a change of speed or by a change in the direction of motion or both, it generates a different type of disturbance. The electromagnetic field now generated varies with time and its effect falls off inversely as the distance from the charge. Such a disturbance is known as 'electromagnetic radiation', and it carries energy away from the charge with the speed of light.

In several astronomical situations, we have huge stores of electric charges in the form of electrons and protons. If some mechanism can be found to accelerate them, we can expect electromagnetic radiation from them. An electric charge responds to a gravitational force, an electric field, and a magnetic field. Of these, the first is usually small, except in the neighbourhood of compact massive objects (see p. 162), and the second almost zero; but the third can be appreciable. A magnetic field cannot change the speed of the charged particle, but it can change the direction of its motion and thus provide acceleration. In Fig. 3.25, we have an electron moving in a circle around a uniform magnetic field perpendicular to the plane of the circle. In practice, of course, the magnetic field is not uniform, either in magnitude or direction; but the general pattern of the electron motion is the same.

As mentioned above, such a motion will generate electromagnetic

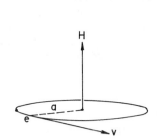

Fig. 3.25. The charge *e* moving in a circle perpendicular to the magnetic field strength **H**, emits radiation mostly in the forward direction of its motion.

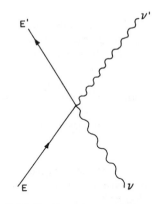

Fig. 3.26. In Compton scattering an electron of energy *E* collides with a photon of frequency ν. After scattering the electron has energy *E′* and the photon has frequency ν′.

radiation. This radiation was first called 'synchrotron radiation' because of its use in a discussion of a laboratory machine called 'a synchrotron' which makes use of this process. Subsequently Alfven and Herlofson applied this process to stars and Ginzburg and Shklovsky to galaxies. Nowadays this process is used in explaining radiation from all sorts of objects, including QSOs, radio sources, X-ray sources, etc., where it is possible to imagine a magnetic field and a source of fast, charged particles.

One important characteristic of synchrotron radiation is that it is polarized, that is, the oscillations of the electric and magnetic fields in the radiation (see Box 2.2) are in a fixed direction rather than being jumbled up. As the other known radiation mechanisms do not have this property, polarized radiation is strongly indicative of a synchrotron source.

The inverse Compton effect

In 1923, A. H. Compton observed that light incident on materials of low atomic weight is absorbed and re-emitted as light of lower frequency. This phenomenon, known as the 'Compton effect', involves the scattering of a photon (that is, a particle of light) by a free electron in the material. Fig. 3.26 shows schematic diagrams for such scattering.

Theoretically, it is not impossible that the scattered photon should have a *higher* frequency than the incident photon. This can happen when a low-frequency photon impinges on a fast-moving electron. The transfer of energy is then from the electron to the photon. Since this is in a reverse direction to that observed in the original Compton effect, this phenomenon is known as the 'inverse Compton effect'.

The inverse Compton effect is particularly useful in astrophysics, where photons of high frequency are required. For example, the process is useful for generating visible light or X-rays. Thus a fast-moving electron in a background of low-frequency radiation, for example, infrared radiation, can produce ultraviolet light or X-rays with the help of the inverse Compton scattering.

Radiation from hot plasma

A plasma is a mixture of ions, which are positively charged, and electrons, which are negatively charged. At high temperatures atoms are mostly stripped of all their electrons, thus creating a plasma. Now in a typical ion–electron collision, the electron is deflected (see Fig. 3.27) and thus suffers a change of velocity. As in the case of

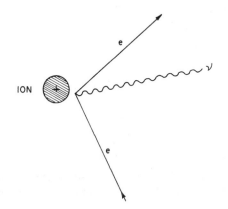

FIG. 3.27. In an ion–electron collision a photon of frequency ν can be emitted.

synchrotron radiation this process, known as *Bremsstrahlung†*, leads to the emission of electromagnetic radiation.

The amount of radiation and the frequency depend on several factors, such as the ion–electron number densities and the electron temperature. The temperature is an indicator of the overall agitation of the plasma. The higher the temperature, the higher is the random velocity of the electrons.

In processes discussed in high-energy astrophysics, this particular method of radiation is important in the generation of X-rays and ultra-violet radiation.

Models of radio sources

The synchrotron mechanism is perhaps the likeliest mechanism for radiation in a typical strong radio source. This is because radio waves are usually polarized. The basic ingredients for synchroton radiation are fast-moving electrons and a magnetic field. Using a simple synchrotron model, it is possible to estimate the total amount of energy in the form of fast electrons and the magnetic field that must

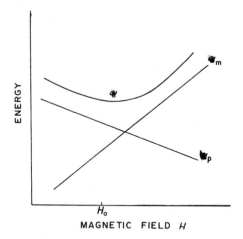

FIG. 3.28. Schematic plot of energy against the magnetic field in the synchrotron model of a radio source. (A logarithmic scale has been used.)

† A German word meaning 'brake-radiation', that is radiation produced by braking the motion of an electric charge.

be present in a typical radio source. This calculation was first perform-ed by Geoffrey Burbidge, in the late 1950s—with startling results. On the basis of 'minimum energy required' a strong radio source like Cygnus A must have a store of energy of the order of 10^{55} J. This estimate (which is about 10^{40} times the energy released in a megaton hydrogen bomb) will have to be pushed *up* if we take into account the fact that the actual situation may not conform with the conditions of the above minimum energy requirement, which demands (1) an overall balance or equipartition, between the energy of electrons and the magnetic field energy, (2) 100 per cent efficiency of the process in operation, and (3) the neglect of the energy of protons or heavy ions. Burbidge's calculation is described briefly in Box 3.5.

---

†Box 3.5 Strong radio sources: energy estimates

Suppose a radio source located beyond our Galaxy sends out a flux (measured on the Earth) of energy l per unit area per unit time in a specific frequency range. The volume V of the emitting region and its distance d from the Earth can be observationally estimated. From this data, can we make an estimate of the total energy reservoir in the radio source? The answer is 'yes', and the calculation is along the following lines.

Usually the mechanism of radiation is the synchrotron process, in which case a magnetic field is involved. For simplicity we will assume that a uniform magnetic field of strength H is present throughout the volume V. Then the energy present in the magnetic field is estimated from the elementary theory of magnetism as

$$U_m = \frac{VH^2}{8\pi}. \tag{1}$$

We also need fast particles in the synchrotron process. Let us suppose that there are N electrons per unit volume, all having the same energy γmc^2, where m is the rest mass of the electron. The factor γ is very large for fast particles, that is, for particles moving with a velocity near the velocity of light (see Box 3.6). In a more elaborate calculation a collection of electrons with various γ values is assumed, but here we will just choose one γ as a representative value. This value is related to the frequency of emitted radiation and to H. Again, we will choose one frequency ν to represent the actual range of frequencies observed. Then the theory of synchrotron emission tells us that

$$\nu = KH\gamma^2, \tag{2}$$

where K is a known constant.

The energy residing in the particles is therefore given by

$$U_{\mathrm{p}} = NV\gamma mc^2 = NV\left(\frac{\nu}{KH}\right)^{\frac{1}{2}} mc^2. \qquad (3)$$

The next step consists of fixing N. Again the synchrotron theory tells us that a single electron with energy γmc^2 in a magnetic field of strength H radiates energy at a rate

$$P = \mu H^2 \gamma^2, \qquad (4)$$

where μ is another known constant. So the total energy radiated by the source is

$$L = NVP = \frac{NV\mu H}{K}\nu. \qquad (5)$$

Arguments given in Chapter 1 then show that the amount of energy received at the Earth per unit area per unit time is

$$l = \frac{NV\mu H\nu}{4\pi K d^2}. \qquad (6)$$

From (3) and (6) we therefore get, by eliminating NV,

$$U_{\mathrm{p}} = \frac{4\pi d^2 l m c^2}{\mu}\left(\frac{K}{\nu H^3}\right)^{\frac{1}{2}}. \qquad (7)$$

Combining (1) and (7) we get the total energy involved as

$$U = U_{\mathrm{m}} + U_{\mathrm{p}} = \frac{V}{8\pi}H^2 + 4\pi d^2 l m c^2\left(\frac{K}{\nu H^3}\right)^{\frac{1}{2}}\mu^{-1}. \qquad (8)$$

In formula (8) we know V, d, l, and ν from observations of a radio source, and we also know m, c, K, and ν from the laboratory physics. The only unknown quantity is H. In Fig. 3.28, U is plotted as a function of H. We see that U has a minimum value U_0 for a certain value H_0 of H. For H greater than or less than H_0, U is higher and keeps on increasing as we go away from H_0 in either direction. So the least energy involved is U_0. For this case

$$U_{\mathrm{m}} = \tfrac{3}{4} U_{\mathrm{p}} = \tfrac{3}{7} U_0. \qquad (9)$$

For values of V, d, l, and ν, given from observations, these estimates come out near 10^{55} J. The magnetic field is around $0\cdot1$ T. (T = tesla, the SI. unit which measures the effect of the magnetic field; $1\,\mathrm{T} = 10^{-4}$ gauss—a gauss is the other common unit of magnetic field strength.)

This calculation and other similar ones which followed presented a serious difficulty for the collision hypothesis about the radiation from a radio source. We have already mentioned the observational problems of the collision hypothesis (p. 79), but a theoretical difficulty arises in the following way. Suppose we have two galaxies of masses M_1 and M_2 in the process of collision. If R is the distance between their centres of mass, the energy of collision is comparable to their gravitational potential energy, which has the magnitude

$$E = \frac{GM_1M_2}{R},$$

where G is the gravitational constant. This result follows from Newton's laws of motion and gravitation, which are supposed to control the motions of galaxies in the above case. If we now set $M_1 = M_2 = 10^{11}M_\odot$, and $R = 50$ kpc, as values typical of galaxies, we get E less than 10^{52} J, that is, less than a thousandth part of the minimum energy required.

The present ideas on what goes on in a radio source are no clearer. Observational evidence indicates some sort of explosion in the nucleus of a galaxy which ejected clouds of fast-moving electrons in two opposite directions. These electrons then emit radio waves under the influence of the ambient magnetic field. While qualitatively this is a reasonable starting point, this picture cannot acccount for all the various structural detail observed in radio sources or in QSOs. Nor do we know anything about the nature of the explosion in the nucleus. Is it an explosion built out of several supernova explosions, somehow triggered off one after another? Does it arise from collisions between densely packed stars? Is this a manifestation of the release of gravitational energy produced by the collapse of a massive object in the nucleus of the galaxy? Or are we witnessing here the primary creation of matter in the Universe? Needless to say, there is no unanimity among astronomers about the answers to these questions.

Quasi-stellar objects as sources of radiation

QSOs seem to emit radiation over a wide range of frequencies. Fig. 3.29 shows a plot of the continuous energy distribution over different frequencies for the source 3C-273. The radiation covers radio, part of the infrared, and optical frequencies, with a gap in the microwave region. The most likely explanation of this radiation seems to lie in the synchroton process and the inverse Compton effect, with the former dominating.

Like strong radio sources, QSOs also pose the problem of the

33 bfrre ffl llll.........

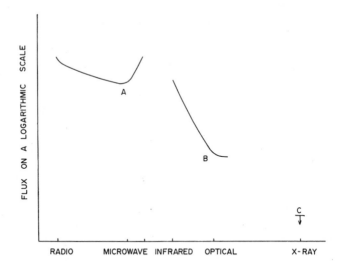

Fig. 3.29. The energy distribution of 3C–273.

primary source of energy. In the case of QSOs the problem is made more difficult by the fact that their distances from us are not unambiguously known. As we saw in Chapter 1, the flux of radiation received at the Earth from a source depends on its distance and on its intrinsic power. For a given flux, the greater the distance the greater is the power, and hence the greater the energy reservoirs required. If distances are related to red-shifts, as required by Hubble's law (see p. 113), then their large red-shifts imply that QSOs are very distant, and consequently their energy reservoirs must be large. If red-shifts are unrelated to distances, QSOs need not be so distant, and their energy reservoirs become that much easier to explain. But a satisfactory QSO model, however, is still not available, whatever we assume about the distance away of a QSO.

X-*ray sources*

X-ray emission from cosmic sources was detected in the 1960s with the help of detectors which were sent up in rockets and satellites. While it was always expected on theoretical grounds that X-ray emitting stars, and even extragalactic sources of X-rays, might exist in the Universe, their actual detection and observation is providing valuable information.

One of the most dramatic sources of X-rays is the Crab Nebula. Continuum X-ray emission from the Crab is believed to be due to the synchrotron mechanism. However, the hot-plasma radiation described above can also play an important part in other X-ray sources.

One such X-ray source is generating considerable excitement at present because of its possible association with a black hole (see Chapters 2 and 5). This is the source Cygnus X-1. It is identified with a supergiant star. However, observations indicate that this star is not alone, but is going round another star with a period of about $5\frac{1}{2}$ days. This other star however, is, *invisible*; but its existence and mass can be inferred indirectly from the observations. It is estimated that the mass is not less than 5–6 solar masses. A star of this mass cannot exist in a stable configuration—it must collapse into a black hole. A black hole is invisible in the sense that it traps all electromagnetic radiation, but it continues to exert a gravitational pull. In the case of Cygnus X-1, this pull attracts plasma from the companion star, and while falling into the black hole this plasma radiates X-rays (see Fig. 3.30).

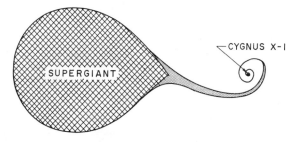

FIG. 3.30. A black hole, shown by a dot, sucks in matter from its companion supergiant star.

As the double star system revolves the supergiant periodically blocks its companion along our line of sight, and this leads to a drop in the X-ray flux. Such a periodic drop is indeed observed in the output of Cygnus X-1.

Cosmic rays

So far we have considered electromagnetic radiation as the main tool of astronomical investigation. We now turn to another tool, which

has become increasingly important over the last two or three decades
—cosmic rays. In the laboratory the physicist is now able to observe
interactions between elementary particles at high energies and for
this purpose he builds accelerators, at enormous expense. These
accelerators can accelerate particles to energies in the range of
10–100 000 MeV (see Box 3.6). It turns out, however, that the

---

†Box 3.6 Particles of high energy

A particle at rest and having a mass m has an equivalent energy
store of magnitude

$$E = mc^2, \qquad (1)$$

where c is the velocity of light. This is one of the important deductions
which form part of Einstein's special theory of relativity.

What does special relativity say about moving particles? If we
observe the above particle in a frame of reference in which it appears
to move with a speed v, it will appear to have a mass γm, where the
γ-factor is given by

$$\gamma = \frac{1}{\left(1 - \dfrac{v^2}{c^2}\right)^{\frac{1}{2}}} \qquad (2)$$

Its energy will accordingly be γmc^2.

According to special relativity, material particles are restricted
to speeds less than c. However, as v approaches closer and closer to c,
the γ-factor increases rapidly. For this reason, particles with velocities
near the velocity of light are considerably more energetic than similar
particles at rest.

Consider the example of a proton. When at rest it has an energy of
nearly 930 MeV. Suppose it is made to move with a velocity which is
less than the velocity of light by the small fraction of one part in two
million. Then (2) gives a γ-factor of nearly 1000. Thus the proton will
now have an energy of 930 000 MeV. The high-energy particles in
man-made accelerators have γ-factors up to about this order of
magnitude. High-energy cosmic rays contain particles with considerably higher γ-factors.

---

atronomer can boast of particles of much higher energies—higher
than the above range by a factor of up to 10^{10}. These particles are
available to him via the cosmic rays which are arriving continuously
from all directions to bombard the Earth. Where do these particles
originate? How are they accelerated? What distances do they cover

before arriving at the Earth? These tantalizing questions have as yet only partly been answered.

How are cosmic rays observed? Direct observations of cosmic–ray particles are possible only by going above the Earth's atmosphere—using balloons, rockets, and satellites. This is because the cosmic rays coming from space interact with the atoms in the atmosphere, so that any ground-based observations would necessarily measure the flux of particles *after* these interactions have taken place. In many cases, and especially at high energies, such indirect methods are the only ones possible, because of the low flux and long measurement time involved. These ground-based methods make use of 'air showers'. These showers contain particles produced after the cosmic rays have interacted with the atoms of the atmosphere. Detectors, ranging in area from several square metres to several square kilometres and placed both at sea level and at mountain altitudes, look for these particles. Very high-energy γ-rays also produce air showers, and their flux is estimated by similar experiments.

Meteorites also provide information about the nature and composition of cosmic rays. A meteorite is a small lump of matter moving in the Solar System. Cosmic-ray particles leave tracks in the meteorite through which they pass. Some particles produce radioactive effects in the solid, and some produce ionization. The tracks can be seen by a chemical process called etching (see Fig. 3.31), and from them it is possible to get information about the cosmic rays as well as about the original matter composition of the meteorite.

What particles do cosmic rays contain? They contain electrons, protons, and the nuclei of various atoms. Protons are the most important component, occurring 10 times more frequently than the heavy nuclei. The iron-group nuclei belong to this last category, and they occur more abundantly in cosmic rays than in ordinary stars. But the contrast between cosmic-ray abundance and the abundance elsewhere in the Universe is much more strongly marked in the case of the lighter nuclei lithium, beryllium, and boron. These occur much more abundantly in cosmic rays than, say, in ordinary stars.

The presence of a mixture of heavy and light nuclei tells us something about the history of cosmic rays. A heavy nucleus travelling at almost the speed of light through interstellar matter will break up into lighter nuclei after a certain period of time. So the longer the cosmic rays travel through the interstellar matter the lower will be the ratio of heavy to light nuclei present in them. The observed ratio

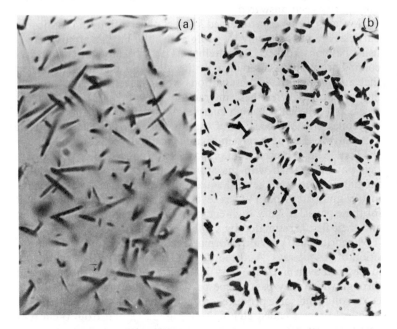

FIG. 3.31. Photomicrographs of fossil tracks, produced by the iron-group nuclei, as seen in feldspar crystals from meteorites found in Moore County (a) and Kapoeta (b). (Photographs by courtesy of D. Lal and his colleagues.)

therefore determines the travel time of cosmic rays through the Galaxy. The time involved is of the order of a few million years (see Box 3.7). So in order to achieve a steady situation, we need to replenish cosmic rays from their sources over this length of time. Where are these sources?

The most likely candidates are supernovae (see p. 42). During a supernova explosion a large flux of high-energy particles is indeed ejected. But a quantitative theory of the supernova origin of cosmic rays has to answer many difficult questions. Are there enough supernovae in the Galaxy to provide an adequate supply of cosmic rays? The answer is 'yes' if we demand a high (but not impossibly high) yield from each supernova (see Box 3.7). The second difficulty arises in trying to explain the isotropic distribution of cosmic rays. If they were produced in the disc of the Galaxy, where most supernovae are likely to be found, we should expect more particles in the plane of the disc than in a direction perpendicular to it, which is not found to be the case. Finally, the most important objection to this theory comes from the fact that cosmic rays of very high energy cannot be confined in the Galaxy. Their pressure is very high—much in excess of the gas pressure and the pressure exerted by the magnetic field. With their excess pressure the cosmic rays would be able to force their way out of the Galaxy and leak into intergalactic space. Thus the very high energy cosmic rays cannot be regarded as lying wholly within the Galaxy.

These difficulties may be lessened if we assume that cosmic rays are not confined to the disc, but leak out into the halo. In the late 1950s Ginzburg and Syrovatsky showed that a viable theory of Galactic cosmic rays can be given provided we choose the various parameters, such as the galactic magnetic field, supernova frequency, and energy output in a supernova explosion suitably (see Box 3.7). But the constraints are so restrictive that they prompt us to consider the alternative suggestion, that cosmic rays are also an extragalactic phenomenon.

Certainly, extragalactic astronomy provides us with a multitude of violently exploding objects like radio galaxies, and QSOs. An order-of-magnitude calculation suggests that there need not be an energy problem (see Box 3.7). The confinement problem does not exist, since in this theory the cosmic rays can freely move in or out of galaxies. But as yet not many theoreticians have seriously explored the pros and cons of this idea of extragalactic cosmic rays.

---

†Box 3.7 The origin of cosmic rays

One clue to the origin of cosmic rays is provided by data on nuclear abundances. The nuclei lithium (Li), beryllium (Be), and boron (B) occur more frequently and the heavier nuclei carbon (C), nitrogen (N), and oxygen (O) occur less frequently in cosmic rays than elsewhere in the Universe. The reason for this is that carbon, nitrogen, and oxygen break up into lithium, beryllium, and boron due to the passage of cosmic rays through the Galaxy, where they collide with ambient matter. The abundance data show that, in their passage through the Galaxy, 1 m^2 of cosmic-ray cross-section should have swept through about 30 kg of ambient matter.

Given the estimated galactic density as about 10^{-21} kg m^{-3} the distance travelled by the cosmic rays must have been about

$$x = \frac{30}{10^{-21}}\,m = 3 \times 10^{22}\,m.$$

Since cosmic-ray particles travel with nearly the velocity of light c ($= 3 \times 10^8$m s^{-1}), the time taken to travel the distance x is

$$T = \frac{x}{c} \cong 3 \times 10^6 \text{ years.}$$

So the cosmic rays we observe now must have been in existence for nearly 3 million years.

The next important clue is provided by the observed energy density of cosmic rays. One cubic metre of space contains 10^{-13} J of cosmic rays. The volume of the disc of the Galaxy is estimated at 10^{61} m^3. So the total energy resident in cosmic rays is

$$E = 10^{61} \times 10^{-13} \text{ J} = 10^{48} \text{ J.}$$

Since the typical lifetime of cosmic rays is about 3×10^6 years, the rate at which new cosmic rays must make up for the decayed ones is estimated at

$$\frac{E}{T} \cong 3 \times 10^{41} \text{ J per year.}$$

Do cosmic rays originate in supernovae? If they do originate mostly in supernovae, we must remember that the energy released in a supernova explosion is around 10^{42}–10^{45} J. If there is one explosion in 30 years, the rate of energy release works out at 3×10^{40}–3×10^{43} J per year. This seems comparable to the above required rate. However, there are difficulties. Supernovae emit predominantly electrons and not as many heavy nuclei as are observed in cosmic rays. Also, the galactic magnetic field, which is no larger than 0·01 T, is not able to control the motion of the very-high-energy cosmic rays. This is seen as follows.

A magnetic field of strength H can hold a proton of rest mass m_p and energy $\gamma m_p c^2$ in an orbit of radius a, where the centrifugal force of circular motion is balanced by the electromagnetic force exerted by H on the charge e of the proton. This leads to the relation

$$\gamma = \frac{eHa}{m_p c^2} \cong 5 \times 10^9.$$

A proton with an energy of up to $\gamma m_p c^2 \cong 5 \times 10^{18}$ eV is therefore within the control of the galactic magnetic field. Cosmic rays seem to contain protons of energy in excess of 10^{20} eV. Such very-high-energy particles cannot be confined within the Galaxy.

This raises the interesting question of whether cosmic rays are a galactic phenomenon or whether they operate on a much grander scale in intergalactic space. The volume of the observable Universe is about 10^{78} m³ (see p. 138). If cosmic rays operate throughout this volume with the observed energy density of 10^{-13} J m⁻³, the total cosmic-ray energy in the observable Universe is 10^{65} J. There are some 3×10^9 galaxies in this region. This means that every galaxy must produce about 3×10^{55} J, over the cosmological time scale of 10 000 million years. We know that radio galaxies do contain bigger energy reservoirs, and so they might play the roles of supernovae in the galactic-origin theory of cosmic rays. Thus from the point of view of the total energy, it is not all that improbable that cosmic rays originate in the galaxies of the Universe.

The question of whether cosmic rays are of Galactic or extra-galactic origin may well have connections with γ-ray astronomy. We have already seen that the inverse Compton effect (p. 90) acts in such a way as to increase the energy of light quanta (or photons). Thus by this process quanta of visible light from the stars may be lifted in energy to become γ-rays, with the help of the high-energy electrons found in the cosmic rays. In extragalactic space the intensity of starlight is fairly high, and if extragalactic cosmic-ray electrons exist they will give rise to a diffuse γ-ray background. Although a similar effect could very well exist in our Galaxy in the case of galactic cosmic rays, its magnitude would be much less than in the extra-galactic case. This point was made by Felten and Morrison. The γ-ray observations to date indicate a background not inconsistent with the extragalactic hypothesis of cosmic rays.

Cosmic-ray electrons play an important part in many ways, although their energy content is only about 1 per cent of the total cosmic-ray energy. This is because high-energy electrons are able to produce radiations of various kinds. We have just seen how they can generate γ-radiation, but they also interact with the magnetic field and produce synchrotron radiation. A typical supernova becomes a radio source through this process. In the Crab Nebula even the X-ray radiation is believed to be of synchrotron origin. Apart from electrons, cosmic rays also appear to contain positrons, but these arise largely through collisions of the heavier particles in cosmic rays with atoms in the Earth's atmosphere.

Conclusion

I have given here a brief survey of the diverse astronomical phenomena occurring in our Galaxy and beyond. Although optical astronomy continues to provide new and interesting information,
astronomy has now been completely transformed because of the data
supplied by the newer branches of science. Indeed, new information
is coming in at such a rate that theoreticians have to continuously
revise their models. It would be premature to say that a consistent
picture has emerged from all the data available so far. The situation
is rather the reverse, but this is only typical of a rapidly growing
subject.

It is worth re-emphasizing here the faith placed by theoretical
astronomers in the basic laws of physics. In the case of stars, we have
seen how the subject of nuclear reactions, a typical laboratory
physics topic, finds application in a stellar interior—a vastly different
situation. A lot of high-energy astrophysics depends on stretching the
laws of electromagnetic theory to limits far beyond those dreamt of
by their originator, James Clerk Maxwell, in the nineteenth century.
Will cosmic rays tell us how the laws of elementary-particle physics
operate at very high energies? Will X-ray astronomy lead us to black
holes? These are exciting prospects for the theoretician willing to
extrapolate his laws to unusual circumstances. But by far the biggest
extrapolation is still to be described—the application of the laws of
physics in constructing models of the Universe—and this forms the
subject of the following chapter.

4

Theories of the Universe

Long before the advent of modern cosmology, the question of the origin and evolution of the Universe fascinated mankind. Different civilizations, religions, and cultures formulated their own interpretations. Of these, the Hindu scripture *Vishnu Purana* comes up with suggestions of a time scale comparable with those of present-day astronomy. In this work (verses 12–15, Book I, Chapter 3), the god Brahma is regarded as the Creator of the Universe, and the overall life of the Universe may be compared with one day of Brahma. One day of Brahma elapses when the four yugas, Kritā, Tretā, Dwāper, and Kali, are repeated a thousand times. One cycle of four yugas takes up 12 000 divine years. One divine year is equal to 360 human years.

Multiplying these factors we see that the day of Brahma is equal to 4320 million years. This may be compared with the estimated age of a big-bang Universe up to the present epoch, which is around 10 000 million years.

In this book we have been progressively working our way towards bigger and bigger astronomical systems. Starting with the stars in our Galaxy we next looked at the galaxies like (or unlike!) our own, and at the other interesting objects in the intergalactic space. The next step brings us to the ultimate: the Universe. The 'Universe' is the largest possible object in existence, and the study of it is not confined to what it looks like 'now' but also includes the investigation of its properties in the past and in the future.

Clearly the scope of our understanding of the Universe is severely limited by many considerations. First, our observing instruments are limited in their ability to observe. Even the best telescope cannot see objects located arbitrarily far away (see Box 4.1). Secondly, our knowledge of the laws of physics, on which we base our explorations of the Universe, is by no means complete. As we shall see in the next two chapters, there is already a good enough case for looking for new

Box 4.1 Astronomical observations: some limitations

Over the last few centuries astronomical techniques have steadily improved, and man has been able to observe more and more aspects of the Universe. Can this process continue endlessly? What limitations are placed by the Universe and the basic laws of science on the observing capacity of the astronomer? To what extent can any such obstacles be surmounted? The following examples may illustrate the present observational situation.

1. One limitation is placed by the sky brightness. A faint object is more readily visible against darker surroundings. It is because city lights increase the background brightness of the sky that the astronomer builds his telescopes in remote regions. However successful the astronomer may be in eliminating man-made sky brightness, he cannot get rid of the brightness produced by the astronomical objects themselves. Although this is small (see p. 185), it is not negligible. And it places a limit on the brightness of an object that can be detected. Thus the 200-inch telescope at Mt. Palomar cannot detect objects fainter than an apparent magnitude of about 22. Since the farther the object the fainter it looks, there is a limit on the distance up to which objects of a particular intrinsic brightness can be seen. For bright galaxies this limit is of the order of 10 000 million light-years. If the overall sky brightness were, say 10 times less than its present value, it would be possible for the astronomer to look much deeper into the Universe, and this might resolve some of the controversial issues. On the other hand, if the sky brightness were 10 times higher than its present value, it would not be possible to observe far beyond our own Galaxy.

2. Another limitation is placed by the fact that light, our usual means of observation, propagates as a wave. For a telescope of a given aperture (that is, diameter of the main receiving lens) the ability to form a sharp image of any object is limited by the wave properties of light. Thus the image of a point source of light is not a point but a small bright circle subtending an angle

$$\theta = \frac{\lambda}{a}$$

at the eye, where λ is the wavelength of light and a is the size of the telescope aperture. This is because of 'diffraction' of the light waves, that is the ability of light to bend slightly round corners. (If light travelled strictly in straight lines the image of a point would be a point.) So the image of an extended object will be slightly blurred. Two points of the object subtending an angle less than θ at the eye will have overlapping images. Hence the full geometrical details of the structure of the source will never be known. The ability to form relatively sharp images is called the 'resolving power' of the telescope, and it depends on the smallness of θ. Because visual wavelengths are considerably shorter than radio wavelengths (see Box 3.4), the optical telescope can achieve a better resolving power with an aperture of a few centimetres than the radio telescope can with an aperture of several metres.

3. The Earth's atmosphere absorbs light of several wavelengths which are important to astronomy. Only in the past decade or so has space technology improved so much as to enable the astronomer to place detectors above the atmosphere and collect information in X-rays, γ-rays, the infrared wavelengths, etc. Naturally the accuracy is not yet so good as that achieved by the ground-based instruments of optical astronomy or radio-astronomy.

ideas in fundamental physics. Finally, the very immensity of the Universe makes it difficult to suppose that the human race is the most superior living species present in it. The biological limitations of the human brain also make it difficult for us to digest information very fast. Even man-made electronic computers can handle information much faster than the best human brain. Thus it is not far-fetched to imagine the existence of other living species, with vastly superior brains, who have a much clearer picture of the Universe than ourselves, just as we have a much better understanding of the Universe than any lower animal species on Earth.

Nevertheless, man must and will continue trying to advance his knowledge. Certainly the factors mentioned above need not and do not deter the scientist from applying his techniques, which have worked reasonably well so far in the case of smaller systems, to the Universe as a whole.

Modern cosmology represents the modern astronomer's attempts to understand the large-scale structure of the Universe. In this chapter we are going to look at some theoretical ideas on the nature of the Universe. But like all other branches of science these astronomical ideas must be subjected to observational tests, and in Chapter 7 we shall return to this confrontation between theory and observation.

Newtonian cosmology

What laws of physics are likely to play an important part in cosmology? Of the basic laws only the law of gravitation seems important enough on a large scale (see p. 141). It is not surprising therefore that early attempts to study the Universe were based on the laws of motion and the law of gravitation, as given by Newton. These attempts, however, failed to provide a satisfactory model. The fault lay partly with the theoretician in attempting to employ the wrong type of model, and partly in the inadequacy of the Newtonian concepts.

Most astronomers of the nineteenth century and those of the early twentieth century believed that the Universe as a whole is static. That is, they assumed that there is no systematic larger-scale motion of its major constituents. So attempts to provide a satisfactory model of the Universe centred round this static picture. However, Newtonian concepts could not give such a model. This is largely because, in a Newtonian framework, every constituent of the Universe attracts every other constituent, with the result that all constituents tend to move *towards* one another, if they are held at rest at any starting moment. If they are all flying *away* from each other at some moment, say as a result of an explosion, they either continue to move away from one another at all times, or they come to a momentary rest and then begin falling in towards one another (Box 4.2). These are the

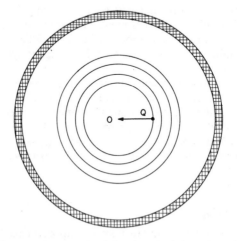

FIG. 4.1.   The matter contained in the shaded region does not exert any gravitational pull on Q.

†Box 4.2 Newtonian cosmology

In Newtonian cosmology the simplest models were obtained by E. A. Milne and W. H. McCrea in 1934, using the cosmological principle and the Weyl postulate. At constant t the Universe is assumed homogeneous and isotropic. So without loss of generality our Galaxy may be taken as the origin of coordinates and the coordinates of a typical galaxy Q of mass m may be denoted by the radial coordinate r and angular coordinates θ and φ. It is shown in Box 4.5 that the only velocity that Q may have relative to our Galaxy, consistent with homogeneity and isotropy, is radially away from the origin and proportional to r. We write this velocity as

$$\mathbf{v} = H\mathbf{r}, \tag{1}$$

where H, Hubble's constant, may depend on time t only. But $\mathbf{v}$ is the rate of change of $\mathbf{r}$, so that (1) reads

$$d\mathbf{r}/dt = H\mathbf{r}. \tag{2}$$

This integrates to

$$\mathbf{r} = S(t)\, \mathbf{r}_0, \; \mathbf{r}_0 = \text{constant}, \tag{3}$$

where

$$dS/dt = H(t)\, S(t). \tag{4}$$

$S(t)$ is the expansion factor. The volume of a sphere of radius r about the origin is

$$V = \frac{4\pi}{3}\, r^3 = \frac{4\pi}{3}\, r_0{}^3\, S^3(t). \tag{5}$$

The density of matter in the Universe may be denoted by the function $\rho(t)$, so that the mass contained in the above sphere is

$$M = V\rho = \frac{4\pi}{3}\, r_0{}^3\, S^3(t)\, \rho(t). \tag{6}$$

As the Universe expands or contracts, if no matter is created or destroyed the mass of this sphere stays constant, that is,

$$\rho(t)\, S^3(t) = \text{constant} = A \text{ (say).} \tag{7}$$

To determine $S(t)$ we note that Q is attracted towards the origin by the gravitational force of matter contained in the sphere of radius r. This force has the magnitude.

$$\frac{GMm}{r^2} = \frac{4\pi}{3}\, \frac{GAr_0 m}{S^2}. \tag{8}$$

Why not include the gravitational effect of other matter in the Universe? This question is answered by dividing this matter into concentric shells of larger and larger radii centred at the origin (see

Fig. 4.1). Newtonian gravitation tells us that the gravitational force exerted by any uniform spherical shell at a particle *inside* it is zero. Hence the above conclusion. In spite of this calculation, the result that Q should be attracted towards the origin appears paradoxical. Surely, as seen from Q, the Universe appears isotropic, and so Q should not be attracted towards any specific point! The paradox arises because we are dealing with an infinite system in an unsatisfactory way, and indicates one of the shortcomings of the Newtonian picture.

Accepting (8) for the moment, we write down the equation of motion of Q as

$$m \frac{\mathrm{d}^2 r}{\mathrm{d}t^2} = - \frac{4\pi r_0 \, AmG}{3 \, S^2},$$

that is, using (3),

$$\frac{\mathrm{d}^2 S}{\mathrm{d}t^2} = - \frac{4\pi AG}{3 \, S^2}, \tag{9}$$

which has the solution

$$\left(\frac{\mathrm{d}S}{\mathrm{d}t} \right)^2 = \frac{8\pi AG}{3S} + B, \tag{10}$$

where B is an arbitrary constant. (9) and (10) show why it is impossible to have a static Universe. In a static Universe, $S =$ constant, so that $\mathrm{d}S/\mathrm{d}t = 0$, $\mathrm{d}^2 S/\mathrm{d}t^2 = 0$. This is not permitted by (9) and (10).

The solutions of (10) can be represented by curves describing the variation of S with time. These curves are very similar to those describing the Friedmann Universe models shown in Fig. 4.9.

predictions of the simplest Newtonian picture, and neither prediction fits with a static Universe. However, as we shall soon see, *the Universe is not static*, and therefore this objection to Newtonian theory is no longer valid.

The second aspect of the matter goes deeper. Even if the Newtonian theory gives a satisfactory model, it is suspect for conceptual reasons. For one of the properties of Newtonian gravitation is that it is an *instantaneous interaction*. That is, if we consider two objects attracting each other gravitationally, say the Sun and the Earth, we have to say that the gravitational influence of one is felt by the other instantaneously. Thus if the Sun were to be suddenly annihilated the Earth would feel the lack of its gravitational pull at once. This is contrary to the modern idea, based on Einstein's special theory of relativity, that no physical influence can travel faster than light. So, in the above example, according to special relativity the Earth should feel the gravitational effect of the Sun's disappearance *at the same time* that the Sun is actually *seen* to disappear from the Earth—*if not later*. The concept of instantaneous interaction does not produce results discrepant with observations in relation to the Solar System because we are dealing with relatively short distances. But its use in cosmology, where distances are measured in thousands of millions of light-years, must be viewed with a great deal of caution. Further, the Newtonian laws of motion have been shown to be incorrect in some circumstances, especially where rapidly moving objects are involved. These laws have now been superseded by those of Einstein's theory of relativity. In cosmology, as we shall see later, we encounter rapidly moving galaxies, and hence their motions are not properly describable by Newtonian laws of motion.

When Einstein proposed his general theory of relativity in 1915, he had a framework which was eminently suited to the discussion of cosmology. This theory took care of the conceptual difficulties discussed above and provided a more sophisticated picture of space and time (see Chapter 5 for a further discussion of general relativity).

In his early attempts to describe the Universe, Einstein tried to construct a static model in line with the attempts of earlier cosmologists. However, his gravitational theory, like Newton's, failed to yield such a model, for more or less the same reasons. To counteract the gravitational attraction, Einstein modified his equations by introducing the now famous 'λ-term'. This new term described a force of repulsion, which became important only over cosmological distances (see Box 4.3). Between this force of repulsion and the force of gravitation the Universe could be held static. This model is known as the 'Einstein Universe', and was proposed by Einstein in 1917.

---

†Box 4.3 The λ-term and the Einstein Universe

In order to produce a model of a static Universe, Einstein introduced a force of repulsion between two masses which increased in proportion to their distance apart. Although this extra force appeared in the equations of general relativity, its effect can be seen in the simpler Newtonian framework described in Box 4.2. There, eqn (9) is changed to

$$\frac{d^2S}{dt^2} = \lambda S - \frac{4\pi AG}{3S^2}.$$

In a static situation the two forces on the right-hand side exactly balance for S given by

$$S_0 = \left(\frac{4\pi AG}{3\lambda}\right)^{\frac{1}{3}}.$$

This is the equivalent of the Einstein Universe in Newtonian cosmology.

The Einstein Universe in fact can be described in the relativistic framework as a Universe with constant $S(=S_0)$ and with curvature parameter $k = -1$. This Universe is *finite* in volume but *unbounded*. To visualize it in lower dimensions, think of the surface of a sphere. It has a finite area, but no boundary.

However, observational astronomy was soon to point a new way along which the theoretical discussion was to proceed, and we shall return to theoretical developments after learning about some events on the observational front.

The expanding Universe

At the beginning of the present century astronomers believed that the entire Universe consisted only of our Galaxy. Within a few years this belief had to be abandoned. New observational criteria (see Box 4.4)

Box 4.4 Matter beyond our Galaxy

Towards the end of the nineteenth century astronomers had detected other galaxies in the form of faint nebulosities; but they assumed that these were part of our Galaxy. The reason for assuming this came from the apparent absence of any such nebulosities in the plane of the Milky Way. At first sight this argument appears sound, for the absence of these objects in the plane of the Milky Way implies some association with the structure of our Galaxy. After all, why should objects well outside our Galaxy avoid this region of the sky?

This view, which persisted in early twentieth century, is now seen to be false. The reason why no nebulosities were seen in the Milky Way plane was because of the existence of absorbing matter in the form of interstellar dust in the disc of the Galaxy. The work of Oort (see Box 3.2) and Lindblad helped greatly in clarifying ideas on the structure of our Galaxy; and when the Galaxy's structure and the presence of interstellar dust were taken into account it became clear that the nebulosities represented objects outside our Galaxy. Hubble's work in the 1920s further demonstrated that these objects are star systems like the Galaxy, and located at enormous distances outside it.

And so ended the monopoly of our Galaxy.

demonstrated unequivocally that certain nebulae which were considered intrinsic to the Galaxy were in fact independent systems
located at enormous distances outside it. Many were subsequently
identified with galaxies like our own, and we have now a vast collection of galaxies of various types in the observable Universe.

In 1928, Edwin Hubble, working on the 100-inch telescope at
Mt. Wilson in Southern California, began to discover a remarkable
property of light from these distant galaxies. This property, known
as the 'red-shift', has been described on p. 74. The shift represents a
systematic increase in the wavelength of light from the source. The
entire spectrum of the source, including the continuum and the
absorption lines, is shifted towards the long-wavelength end, that is,
towards the red end.

In the case of galaxies, the main feature in the spectrum is the
absorption lines. These lines correspond to absorption processes
taking place in the atoms and molecules of the galaxies. From our
own Galaxy, and from some nearby ones, we know at what wavelengths to expect these lines. In Hubble's observations, the lines from
distant galaxies appeared at longer wavelengths, compared to the
lines of our own Galaxy or of nearby ones (see Fig. 4.2). Thus if we
expect the line to be of wavelength λ_o and it appears instead to have a
wavelength λ (greater than λ_o) then we define the red-shift z by

$$z = \frac{\lambda - \lambda_o}{\lambda_o}. \tag{4.1}$$

Hubble found such red-shifts in most of the distant galaxies.
Moreover, he noticed that the fainter the galaxy the larger is its
red-shift (see Fig. 4.3). If we take the view that the faintness is
entirely due to distance, then we come to the conclusion that z
increases with distance. Hubble found a simpler linear law relating z
to the distance D:

$$z = \frac{DH}{c}, \tag{4.2}$$

where H is a constant and c is the velocity of light. Hubble estimated
the value of H as approximately $1 \cdot 5 \times 10^{-17}$ per second. This constant,
known as Hubble's constant, has been measured and re-measured by
astronomers over the last 40 years or so. Each time its value has been
reduced, the present value being $1 \cdot 5 \times 10^{-18}$ per second, i.e. 10 times
smaller than Hubble's original estimate.

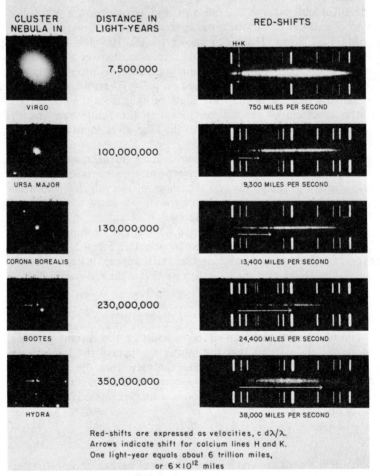

RELATION BETWEEN RED-SHIFT AND DISTANCE FOR EXTRAGALACTIC NEBULAE

CLUSTER NEBULA IN	DISTANCE IN LIGHT-YEARS	RED-SHIFTS
VIRGO	7,500,000	750 MILES PER SECOND
URSA MAJOR	100,000,000	9,300 MILES PER SECOND
CORONA BOREALIS	130,000,000	13,400 MILES PER SECOND
BOOTES	230,000,000	24,400 MILES PER SECOND
HYDRA	350,000,000	38,000 MILES PER SECOND

Red-shifts are expressed as velocities, $c\,d\lambda/\lambda$.
Arrows indicate shift for calcium lines H and K.
One light-year equals about 6 trillion miles,
or 6×10^{12} miles

FIG. 4.2. Relation between red-shift (velocity) and distance of galaxies: direct and spectrum photographs of some nearby extragalactic nebulae. Velocities are in miles per second and distances in light-years. (Photograph from the Hale Observatories.)

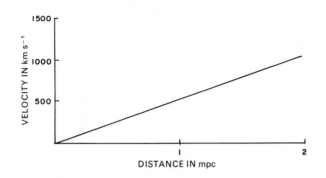

FIG. 4.3. Hubble's velocity–distance relation. The velocity here is c times the red-shift.

The Doppler effect

How can we interpret the cause of the red-shift? The simplest interpretation lies in the 'Doppler effect', In Fig. 4.4(a) we have a source S and a receiver R of light, at rest relative to each other. Suppose at time $t = 0$ S sends out its first wave to R. If R is located at a distance D from S, the wave, travelling with velocity c, will reach R at a time $t = D/c$. If ν_0 is the frequency of the radiation and λ_0 its wavelength, the next wave will be sent by S at the time $1/\nu_0$, and will reach R at the time $t = D/c + 1/\nu_0$. R will therefore interpret ν_0 to be the frequency of the wave.

Suppose now that S is moving away from R with velocity v and that Fig. 4.4(a) represents the situation at $t = 0$, when the first wave left S. As before, it will be received by R at $t = D/c$. The next wave will be sent by S at a time $t = 1/\nu_0$ when the distance between R and S will have increased to $D + v/\nu_0$ (see Fig. 4.4(b)). So this wave will reach R at a time

$$t = \frac{1}{\nu_0} + \frac{D}{c} + \frac{v}{c\nu_0} = \frac{D}{c} + \frac{1}{\nu_0}\left(1 + \frac{v}{c}\right)$$

Therefore R will interpret the frequency of the wave as ν where

$$\frac{1}{\nu} = \frac{1}{\nu_0}\left(1 + \frac{v}{c}\right)$$

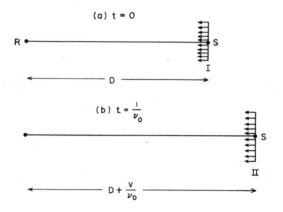

FIG. 4.4.

and its wavelength as

$$\lambda = \lambda_0 \left(1 + \frac{v}{c}\right).$$ (4.3)

Hence R encounters a red-shift given by

$$z = \frac{v}{c}.$$ (4.4)

This is the Doppler effect, and it applies to any wave motion. In the case of sound waves the drop in frequency of a receding source shows itself in the flattening of pitch. Similarly, when S is approaching R, v is greater than v_0, and the sound appears shriller. In the case of light, the spectrum from an approaching object is blue-shifted.

We obtained the above formula on the basis of classical Newtonian physics. It has to be modified to take into account the effect of special relativity when v is close to the velocity of light. As seen by R, the clock of S appears to go slow by a factor $(1 - v^2/c^2)^{\frac{1}{2}}$, where c is the velocity of light. So applying the above example to light waves, the frequency ratio as measured by R and S respectively is changed to

$$\frac{v}{v_0} = \frac{\left(1 - \dfrac{v^2}{c^2}\right)^{\frac{1}{2}}}{\left(1 + \dfrac{v}{c}\right)} = \frac{\left(1 - \dfrac{v}{c}\right)^{\frac{1}{2}}}{\left(1 + \dfrac{v}{c}\right)^{\frac{1}{2}}},$$

and we get

$$1+z = \frac{\left(1+\dfrac{v}{c}\right)^{\frac{1}{2}}}{\left(1-\dfrac{v}{c}\right)^{\frac{1}{2}}}. \tag{4.5}$$

Thus z tends to infinity when v approaches c. When z becomes comparable with 1 we should use the relativistic formula.

The recession of galaxies

If we interpret the galactic red-shift as due to the Doppler effect we can use the simple Newtonian relation (4.4) since the red-shifts measured by Hubble were small compared with 1. Combining this relation with Hubble's observed relation (2) we get

$$v = HD.$$

That is, the further a galaxy is from us the faster (in proportion) is the velocity of recession of that galaxy.

This might give the impression that we are in a singularly unique position with respect to the rest of the Universe, with every other galaxy receding from us. However, this is not the case. If we observed the Universe from any other galaxy we would notice exactly the same phenomenon. This result is at first difficult to believe, but it follows from simple arguments, as explained in Box 4.5.

---

†Box 4.5 The velocity–distance relation

The velocity–distance relation

$$\mathbf{v} = H\mathbf{D} \tag{1}$$

is consistent with the cosmological principle. In (1) we have used vectors to indicate magnitude and direction, so that the equation not only shows that velocity is proportional to distance but also tells us that the velocity vector is directed radially away from our Galaxy.

Suppose observations are made from another galaxy G at distance vector $\mathbf{r}$. The velocity of G relative to our Galaxy is given by (see Fig. 4.5)

$$\mathbf{w} = H\mathbf{r}. \tag{2}$$

A third galaxy G′ is located at $\mathbf{D} = \mathbf{r}'$ with respect to our Galaxy. This has velocity relative to our Galaxy given by

$$\mathbf{w}' = H\mathbf{r}'. \tag{3}$$

The velocity of G′ relative to G is therefore

$$\mathbf{v}' = \mathbf{w}' - \mathbf{w} = H(\mathbf{r}' - \mathbf{r}) = H\,\mathbf{D}', \qquad (4)$$

where $\mathbf{D}' = \mathbf{r}' - \mathbf{r}$ is the distance vector of G′ relative to G. So (4) is the velocity–distance relation measured by an observer on G. This has the same structure as (1). This is consistent with the homogeneity requirement of the cosmological principle. Since (1) is the same in all directions, it is also consistent with isotropy. In fact it can be shown (see *Cosmology* by H. Bondi, Cambridge University Press, 1966) that (1) is the most general non-trivial relation between $\mathbf{v}$ and $\mathbf{D}$ consistent with the cosmological principle.

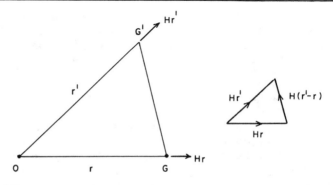

FIG. 4.5.

To visualize the exact situation we can follow the 'balloon analogy'. If we blow up a balloon with dots on its surface, we notice all dots moving away from one another. Yet no particular dot has any favoured position in relation to the rest. Another analogy is that of a cubical lattice of metal bars. If the lattice is heated, all the bars expand, and the vertices of the lattice move away from one another. From these analogies we form the following picture of the Universe. We imagine the galaxies as embedded in a space which is itself expanding. Thus all galaxies move away from one another, leading to the observed Doppler effect. This picture is often described by the phrase 'the expanding Universe'.

The cosmological postulates

We now return to the theoretical aspects of the cosmological problem, in particular to the question, 'How can we construct mathematical models of the Universe?' Because of its sounder foundations, theoreticians have preferred Einstein's general theory of relativity to Newton's laws of motion and gravitation as a framework in which to describe the Universe. Einstein's theory, however, is very complex and does not easily yield exact mathematical solutions. So rather than attempt to construct very general models, cosmologists have gone in for special types of model based on a couple of simplifying assumptions. These two assumptions are called the 'Weyl postulate' and the 'cosmological principle'. It is indeed fortunate that sufficient regularity exists in the Universe to permit the use of these assumptions.

1. *The Weyl postulate.* The Universe in principle could be a very complicated system, with various astronomical objects moving in it in a jumbled kind of way (see Fig. 4.6(a)). However, in practice it is

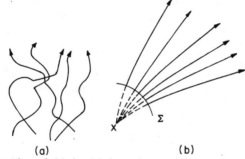

(a)   (b)

Fig. 4.6. Examples of (a) jumbled motion and (b) motion according to Weyl's postulate.

possible to visualize the simpler picture shown in Fig. 4.6(b), in which we have galaxies streaming along regular tracks which do not intersect each other. The galaxy tracks, in either case, are drawn in a space–time diagram. The significance of the point X where all tracks appear to meet will be discussed later.

However, the 'time' in Fig. 4.6(a) has no great physical significance. Einstein's relativity theory tells us that each galaxy will have its own time-keeping device, and in a jumbled-up case of this type it would not be possible to synchronize clocks on different galaxies. In Fig. 4.6(b) such a synchronization is possible thanks to the regularity of motions involved. Here we can draw a surface Σ intersecting all tracks at right-angles, and argue that at all such points of intersection the clocks record the same time. In this way we have a

universal or a 'cosmic' time which serves as a reference coordinate
for the Universe as a whole. The Weyl postulate gives a precise
mathematical definition of this regularity. With its help the large-
scale structure of the Universe is considerably simplified. Moreover,
the cosmologist can speak of a time coordinate or an 'epoch' with
respect to which any changes in the Universe can be described.

2. *The cosmological principle*. This introduces a further simplifica-
tion, especially in relation to the structure of the surface Σ mentioned
above. It states that a typical Σ surface is *homogeneous* and *isotropic*.

In physical terms this can be interpreted as follows. Suppose we fix
the value of the cosmic time t and look at the Universe from any of
the galaxies which are moving according to the Weyl postulate. Then
the Universe will look the same—no matter which galaxy we choose
to observe it from. Moreover, it will present the same view in all
directions from any galaxy. To put it another way, if you were taken
blindfolded and left in any part of the Universe, upon opening your
eyes you could not tell where you are or in which direction you were
looking.

With the help of these two assumptions the cosmologists are able
to simplify the overall problem considerably. In essence the structure
of the Universe may be described in terms of two physical quantities.
One is the 'sign of curvature' of the Σ-spaces. That is, we may imagine
the Σ-spaces as spaces of positive, negative, or zero curvature (see

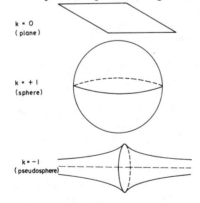

k = 0
(plane)

k = + 1
(sphere)

k = -1
(pseudosphere)

FIG. 4.7.   Two-dimensional examples of surfaces of constant curvature.
In cosmological spaces we have to imagine *three*-dimensional
surfaces of constant curvature.

Fig. 4.7). We may attach values 1, -1, or 0 to these three possibilities. The second parameter is the 'expansion factor'. We may denote it by a function $S(t)$ of the cosmic time t. In physical terms it describes how the separation between any two galaxies, measured at any cosmic time, changes with the cosmic time. If $S(t)$ increases with t, it means that the Universe is expanding, since any two galaxies are moving further apart; if $S(t)$ decreases with t, the Universe contracts; whereas if $S(t)$ stays unchanged, the Universe remains static.

The cosmological red-shift

The function $S(t)$ is related to the phenomenon of the observed spectral shift. Suppose a light wave is sent from a distant galaxy G at cosmic time t_1 and is received by us at the cosmic time t_2. Suppose the next wave is sent out by G at a time $t_1 + 1/\nu_o$, where ν_o is the frequency of the wave measured at G. When will it arrive at our Galaxy? Not at time $t_2 + 1/\nu_o$, but at a time $t_2 + 1/\nu$, where

$$\frac{\nu}{\nu_o} = \frac{S(t_1)}{S(t_2)}.$$

This curious effect arises from the property of light propagation in non-Euclidean geometry (see Chapter 5). Since in the expanding Universe $S(t_2) > S(t_1)$, the waves will be red-shifted (see p. 123 for further details).

Although we introduced the concept of an expanding Universe through the idea of the Doppler shift, the above interpretation, based strictly on general relativity, is more rigorously correct. Once we adopt general relativity as our framework for describing the Universe, we have to think in terms of non-Euclidean geometry and here the concept of velocity, which was used in our Doppler-shift evidence for an expanding Universe, is somewhat ambiguous. The above explanation does not suffer from these defects and is often designated by the name 'cosmological red-shift'.

To decide the actual behaviour of $S(t)$ and to know whether the Σ-surfaces have positive, negative, or zero curvature, the cosmologists resort to Einstein's equations, and this problem will be discussed in the following section.

The Friedmann models

In the 1920s, even before the concept of the expanding Universe became established, theoreticians were exploring the cosmological

solutions of Einstein's equations. Einstein had obtained a static
solution of his equations by including a long-range repulsive force,
and it was his hope that this solution would describe the overall
structure of the Universe in a simplified way. In his model the
'curvature' of space is related to the density of matter filling up the
space, and this emphasizes the central theme of the general theory of
relativity that matter distribution and space–time geometry are
related in an intimate fashion. It would be nice if this had turned out
to be the only simple solution of the equations; but it was not to be.

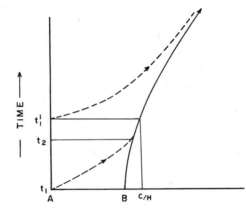

FIG. 4.8.   Radial proper distance of B from A.

Subsequently, de Sitter found an alternative and equally simple
solution in which the Universe is empty but expanding (see Box 4.6).

---

†Box 4.6 The de Sitter Universe

In the de Sitter Universe, the parameter $k = 0$ and the expansion
function is given by

$$S = \exp Ht,$$

where H is Hubble's constant, which does not depend on time. It is, in
fact, given in terms of the λ-repulsion force by

$$H^2 = \tfrac{1}{3} \lambda.$$

There is no matter in the de Sitter Universe, but the distance between
test-particles, following Weyl's postulate, increases in the proportion
of the expansion factor S. This Universe has a so-called 'event

horizon' which arises in the following way. Consider two test-particles A and B moving away from each other in the manner mentioned above. Normally, a light signal sent by A at a given cosmic time t_1 will reach B at a later cosmic time t_2. But a stage will come when B will have moved so far that the signal sent at t_1' will *never* reach B. When this stage comes B is said to have crossed beyond the event horizon of A. This happens when B's distance from A, as measured at time t_1', is c/H, where c is the velocity of light (see Fig. 4.8).

That is, if we imagine any two observers, who are like the galaxies moving according to the Weyl postulate, these observers will be receding from each other. The de Sitter Universe therefore has 'motion without matter', in contrast to Einstein's Universe which is characterized by 'matter without motion'.

Both these models require the λ-force of repulsion. However, by 1922, Friedmann had obtained expanding models from Einstein's equations *without such a force*. The early work of Friedmann has since been further developed by a number of theoreticians. All the Friedmann-type models are based on the Weyl postulate and the cosmological principle. The Einstein equations relate the behaviour of these models, as characterized by the expansion factor S and the curvature parameter, to the distribution of matter and energy in the Universe (see Box 4.7 for details). Briefly, the solutions can be

---

†Box 4.7 The Friedmann models

These are obtained by solving Einstein's equations under the simplifying conditions of the Weyl postulate and the cosmological principle. Assuming that matter in the Universe has a density $\rho(t)$ at cosmic time t, Einstein's equations reduce essentially to the following two equations:

$$\rho S^3 = \text{constant} = A \text{ (say)}, \tag{1}$$

$$\left(\frac{dS}{dt}\right)^2 + kc^2 = \frac{8\pi GA}{3S}. \tag{2}$$

The similarity between (1) and (2) and the Newtonian equations of Box 4.2 is remarkable; but these equations are free from the somewhat ambiguous arguments of Newtonian cosmology. The constant k has values $+1$, -1, or 0. The solutions for S as a function of t can be obtained in all three cases. These are illustrated by curves in Fig. 4.9. In the cases $k = 0$, -1, the expansion factor increases steadily from zero to infinity. In the case $k = +1$ it reaches a maximum value and decreases back to zero.

The model with parameter $k = 0$ is known as the Einstein–de Sitter model. The expansion factor here is given by

$$S(t) = (6\pi GA)^{\frac{1}{3}} t^{\frac{2}{3}} \tag{3}$$

(note that this is different from the Einstein Universe or the de Sitter Universe referred to earlier in the text).

As in the Newtonian case, Hubble's constant is given by

$$H(t) = \frac{\mathrm{d}S}{S\,\mathrm{d}t}. \tag{4}$$

In the Einstein–de Sitter model we get $H(t) = 2/3t$. Thus if H_0 is the present value of Hubble's constant, the age of the Universe is given by

$$t = 2/(3H_0) \cong 13 \times 10^9 \text{ years.} \tag{5}$$

See Chapter 7 for a further discussion of the 'age' problem.

described with the help of Fig. 4.9, where we have plots of the function $S(t)$ against the cosmic time t. There are three curves corresponding to positive, zero, or negative curvature of the Σ-surfaces; but all three start from the point $S = 0$, $t = 0$. In the case of zero or negative curvature ($k = 0, -1$), the function $S(t)$ continually increases with t. This represents an ever-expanding Universe. In the case of positive

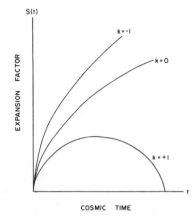

FIG. 4.9. Plots of $S(t)$ versus t for three different Friedmann models of the Universe.

curvature ($k = +1$), S increases up to a certain maximum value, and then decreases again. This represents a Universe which expands up to a certain stage and then contracts again until $S = 0$ is reached. It is tempting to argue that such a model describes a pulsating Universe. This is not correct. In a pulsating model the expansion and contraction phases should alternate. In the present case contraction follows expansion but not vice versa. Indeed, within the framework of Einstein's general relativity it has not been possible to demonstrate that a contracting Universe, on reaching the stage $S = 0$, will re-expand.

What does $S = 0$ mean? Mathematically speaking, this corresponds to a 'singularity'. At $S = 0$ the mathematical equations involved break down because certain quantities become zero or infinite and cannot be given precise meanings. Physically speaking, the density of matter and energy at $S = 0$ are infinite and so are the velocities of matter particles. This state of the Universe is often characterized by the concept of the big bang. At $S = 0$, the tracks of all galaxies in the Weyl postulate meet at one point. This is the point X of Fig. 4.6(b).

So we have the following description of a big-bang Universe. At an epoch, which we may denote by $t = 0$, the Universe explodes into existence. The debris of the explosion flies apart at tremendous speeds. The gravitational pull between the different constituents tries to put a brake on this motion, and it succeeds in stopping the expansion and changing it to contraction only if the Σ-surfaces have positive curvature, that is, if the parameter $k = +1$. In cases where the curvature is negative or zero, the expansion tends to be slowed down but never brought to a complete stop.

The epoch $t = 0$ is taken as the event of 'creation'. Prior to this there existed no Universe, no observers, no physical laws. Everything suddenly appeared at $t = 0$. The 'age' of the Universe is defined as the cosmic time which has elapsed since this event, and is estimated at around 10 000–15 000 million years (see Chapter 7).

Although scientists are not in the habit of discussing the creation event or the situation prior to it, a lot of research has gone into the discussion of what the Universe was like immediately after its creation. In the following section I shall describe the pioneering work of George Gamow in connection with the origin and synthesis of chemical elements; but apart from this work, Gamow also conjectured that the early Universe was largely made of high-intensity radiation rather than matter. (see Box 4.8). In this theory, subsequent expansion of the Universe reduced the intensity of radiation much more rapidly than it

---

†Box 4.8 The radiation Universe

We saw in Boxes 4.2 and 4.7 that as the Universe expands the density of matter falls as

$$\rho = \frac{A}{S^3}, \quad A = \text{constant.} \tag{1}$$

If instead of matter we were considering a mixture of matter and electromagnetic radiation, we would have found that u, the density of radiation, diminishes with expansion according to the formula

$$u = \frac{B}{S^4}, \quad B = \text{constant,} \tag{2}$$

provided there is no interchange between matter and radiation in the course of expansion. The matter density ρ continues to satisfy the relation (1). So at any epoch we have

$$\frac{\rho}{u} = \frac{AS}{B}. \tag{3}$$

In a big-bang Universe S starts from zero, at which stage, and for a period after it, ρ is small in comparison with u. This leads us to the concept of the radiation–dominated Universe. For example, in considering the Einstein–de Sitter model in Box 4.7 we took eqn (1) as the starting point. If radiation is also present, and is the more dominant of the two, close to the big-bang epoch, we should use (2) instead of (1). The corresponding equation ($k = 0$) then becomes

$$\dot{S}^2 = \frac{8\pi GB}{3S^2c^2}. \tag{4}$$

This has the solution

$$S = \left(\frac{32\pi GB}{3c^2} \right)^{\frac{1}{4}} t^{\frac{1}{2}}. \tag{5}$$

Therefore the density of radiation at a time t is given by

$$u = \frac{B}{S^4} = \frac{3c^2.}{(32\pi G)\, t^2}. \tag{6}$$

Electromagnetic radiation of this form is believed to follow the black-body law (see Box 2.3). Accordingly, its temperature is given by

$$T = \left(\frac{u}{a} \right)^{\frac{1}{4}} = \left(\frac{3c^2}{32\pi Ga} \right)^{\frac{1}{4}} t^{-\frac{1}{2}}. \tag{7}$$

That is, T drops off as the reciprocal of $t^{\frac{1}{2}}$. Note that all the constants in (7) are known from laboratory physics: c is the velocity of light, G is the gravitational constant, and a is the radiation constant. So from (7) we can tell the temperature of the Universe a few moments after the big-bang. For example, one second after the big-bang the temperature is

$$\left(\frac{3c^2}{32\pi Ga} \right)^{\frac{1}{4}} \simeq 15\ 000 \text{ million K } (1 \cdot 5 \times 10^{10} \text{ K}). \tag{8}$$

reduced the density of matter, so that the present state of the Universe is matter-dominated. According to Gamow a very faint background of radiation—the relic of the big bang—should also exist at the present epoch. There was great excitement, therefore, when in 1965, Arno Penzias and Robert Wilson at the Bell Telephone Laboratory reported an isotropic radiation background in the microwave region (we shall return to this background radiation in Chapter 8).

Apart from these models already mentioned, other big-bang models based on the repulsive force introduced by Einstein have been widely investigated. Pioneering work on such models was done by Abbé Lemaitre and by Sir Arther Eddington. (A brief description of the Eddington–Lemaitre model is given in Box 4.9). The need for the repulsive force, as mentioned before, arose because Einstein wanted to construct a static model. However, when it was known that the Universe is not static, the overriding need for this force disappeared,

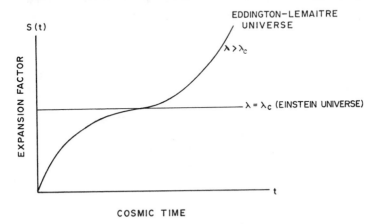

FIG. 4.10.

and Einstein himself abandoned it. Today a majority of cosmologists prefer to do without the repulsive force in their models, although there is no direct observational evidence for or against it. This preference is largely due to the theoretician's desire to have as simple a theory as possible. He goes for more elaborate theories only if forced into them by observational evidence.

---

†Box 4.9 The Eddington–Lemaitre model

Eddington and Lemaitre took the λ-term of Einstein seriously, and constructed cosmological models taking this into account. The result is mainly to modify eqn (2) of Box 4.7 to

$$\left(\frac{dS}{dt}\right)^2 + kc^2 - \lambda S^2 = \frac{8\pi GA}{3S}. \tag{1}$$

Fig. 4.10 shows the general behaviour of $S(t)$ for various values of λ, in the more interesting case of $k = 1$.

There is a critical value $\lambda = \lambda_c$ when (1) admits a static solution. This is the Einstein Universe discussed in Box 4.3. For $\lambda > \lambda_c$ the Universe starts with a big-bang, slows down for a period, and then expands to infinity. For $\lambda < \lambda_c$ the Universe starts with a big-bang and contracts back to the state where $S = 0$. The Einstein Universe itself is somewhat precariously poised between these two cases. It is unstable and upon slight disturbance it either contracts or expands. Eddington and Lemaitre favoured the latter possibility, in which the Universe, after spending a considerable time in the near-static Einstein state, expands to infinity. They believed, for example, that galaxy formation took place in the static phase.

Element synthesis in the hot big bang

In the mid-1940s George Gamow suggested that the high density and high temperature required for the synthesis of elements existed in the few moments after the big bang Universe exploded into existence. In his simplified picture, Gamow supposed the Universe to be made initially of neutrons and photons. Neutrons are charge-free particles found in atomic nuclei, while photons are the carriers of light (or light quanta). In this case, the temperature T of the Universe, t seconds after its origin, is given by

$$T = 1\cdot5 \times 10^{10}\, t^{-\frac{1}{2}}\ \text{K} \tag{4.6}$$

(for details see Box 4.8). Thus the one-second-old Universe has a temperature of 15 000 million K (15×10^9 K).

In normal circumstances the neutron is not a stable particle. It decays into an electron, a proton, and an antineutrino in about 700 s. More precisely, of a given population of neutrons only half will remain intact after 700 s. A universe which is 700 s old will have a temperature of about 500 million K (5×10^8 K). At this stage we will have roughly equal numbers of neutrons and protons. If the matter density is sufficiently high, the neutron and the proton will be fused together to form deuterium (heavy hydrogen). Two deuterium atoms will then combine to form helium, in the manner described on p. 35. How much helium is formed depends very sensitively on how dense the matter is. Unless the matter density is carefully adjusted either no helium is formed or all matter is converted to helium. Gamow estimated a relation of the form

$$N \cong 10^{-10}\, T^3 \qquad (4.7)$$

between the number N of nucleons and the temperature T, in order to produce approximately equal quantities of helium and hydrogen.

In 1950 the Japanese astrophysicist Hayashi pointed out an important aspect which greatly modified these considerations. Initially, when the electromagnetic radiation is powerful, electron–positron pairs are produced at the expense of photons. The positrons help in forming protons with the reaction

$$n + e^+ \rightarrow p + \nu.$$

Thus at the very high temperature right at the beginning of the Universe all such particles, which include neutrons, protons, electron–positron pairs, and neutrino–antineutrino pairs, will be in what is known as 'thermal equilibrium'. Thermal equilibrium represents the state of maximum likelihood for the system. That is, it corresponds to the most likely distribution of the various species of particles involved. The laws of statistical mechanics determine the proportions of the different particles in a thermal equilibrium. One result of this is that the number of protons is greater than the number of neutrons. However, as the Universe expands, a stage comes when the reaction time becomes smaller than the characteristic time of expansion of the Universe, and equilibrium no longer operates. Then the number of neutrons decreases less rapidly than that required by thermal equilibirum.

The presence of so many particles effectively enhances the radiation field, and the Universe, as a result, expands more rapidly. Instead of the Gamow relation (4.6), we how have

$$T = 10^{10}\, t^{-\frac{1}{2}}\ \text{K}. \qquad (4.8)$$

Apart from this difference, the other difference shows itself in helium production. The production of helium is no longer so sensitively dependent on N. For a reasonably wide range of matter densities, roughly the same proportion of helium is produced. This is shown by the plateau in Fig. 4.11. The plateau, according to present estimates

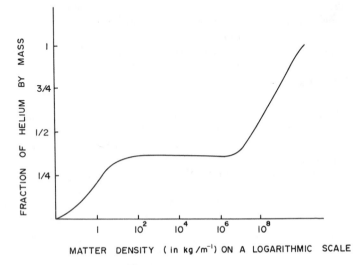

FIG. 4.11. Helium synthesis in the big bang.

of nuclear reactions, is at 35 per cent or so. Is there so much helium in the Universe? Observational estimates as yet do not give a clear-cut answer to this question.

There is another check on this calculation. The electromagnetic radiation present at helium formation is subsequently decoupled from the matter and cools down. Its temperature in the present-day epoch can be estimated and compared with observation. The fact that such radiation is indeed observed is taken by many astronomers as an indication that Gamow's original concept of a hot-radiation Universe with a big bang origin is right (and I shall return to this point in Chapter 7).

Although a plausible theory of helium generation can be made out along these lines, the same ideas do not help towards a theory of the synthesis of heavier elements. For instance, we cannot explain how carbon or oxygen could be formed in this way. For these elements, in any case, we must go back to our other source of nucleosynthesis, the star.

The steady-state theory

In 1948 three British astronomers, Hermann Bondi, Thomas Gold, and Fred Hoyle, put forward an alternative to the big-bang model of the Universe. This alternative theory is commonly known as the 'steady-state theory'. Because of its radical outlook this theory has stirred up considerable interest among astronomers and physicists, and attitudes to it have varied from enthusiasm to hostility. For the present, the salient points which characterize the theory will be described, leaving its merits to be discussed in Chapter 7.

The approach of Bondi and Gold to the steady-state theory is different from that of Hoyle, although the overall picture which finally emerges is the same. Bondi and Gold modified one of the cosmological assumptions to arrive at their theory, whereas Hoyle modified Einstein's equations. I shall describe both approaches in turn.

The perfect cosmological principle

We have already seen that the cosmological principle requires the Universe to be homogeneous and isotropic *at a given cosmic time*. This however, still permits a variation from one Σ-surface to another; that is, the Universe is allowed to change with time. Bondi and Gold modified this statement to say that the Universe looks the same at all cosmic times—in addition to what is required by the ordinary cosmological principle. The new principle, which guarantees uniformity of structure not only in space but also in time, was called the 'perfect cosmological principle' (PCP) by its authors.

Apart from the fact that the PCP restores time to the same footing as space and is thus intellectually more appealing, it also has a direct bearing on the observational interpretation of the structure of the Universe. When we look at a distant galaxy, we do not see it as it is at present but as it was a long time in the past. This is because the light we are receiving from it travels with a finite speed and takes some time to reach us. Now, the usual procedure in cosmology consists of making comparisons between the distant and the nearby parts of the Universe. For instance, Hubble's law is based on the premise that the wavelength of light emerging from an atom in a distant galaxy does not differ from that coming from a similar atom in the laboratory. The red-shift arises because this light undergoes an increase in wavelength during its transit through an expanding space. This interpretation is based therefore on the tacit assumption that the laws of physics which operated in the remote past in the distant galaxy (which determine the wavelength of light emitted) are the same as those which operate now.

Bondi and Gold questioned this assumption in the context of a big-bang Universe. They argued that such a Universe evolves with time, and so conditions in the past were not the same as they are now. For instance, just a second after the big bang the Universe had an estimated temperature of 10^{10} K, far above what is observed now. If the Universe is supposed to include everything, it must also somehow contain the laws of physics; and in a changing Universe, how can we guarantee that the laws of physics have remained unchanged? We are, of course, free to make such an assumption, but the foundations on which it is based are shaky, and require the physical laws to have an eternal status immune from the influence of the Universe. The only situation in which such an assumption is firmly based is in the case of a Universe *which does not change with time*. Hence the need for Bondi and Gold's PCP. This unchanging Universe is called the 'steady-state Universe'.

With the help of the PCP, Bondi and Gold were able to deduce a number of important properties of the steady-state Universe. As we shall see in the next chapter, they were able to deduce that such a Universe must continually expand.

However, as the Universe expands, its existing galaxies must move apart from one another. If we look at the Universe now and, say, 10 000 million years later, we would notice a thinning out of the material in the Universe. But this is contrary to the PCP, which requires the material density to remain the same. To reconcile this we need new material to appear in the Universe to make up for the depletion produced by the expansion (see Fig. 4.12). This steady or continuous creation of matter has been one of the main points attacked by the critics of the steady-state theory.

Where does the matter come from? Does it represent a violation of the law of conservation of matter and energy? Such questions were not fully answered by Bondi and Gold, who argued essentially that once we accept the PCP we must accept all its consequences without question. While this is a deductive approach, it is unfamiliar to the

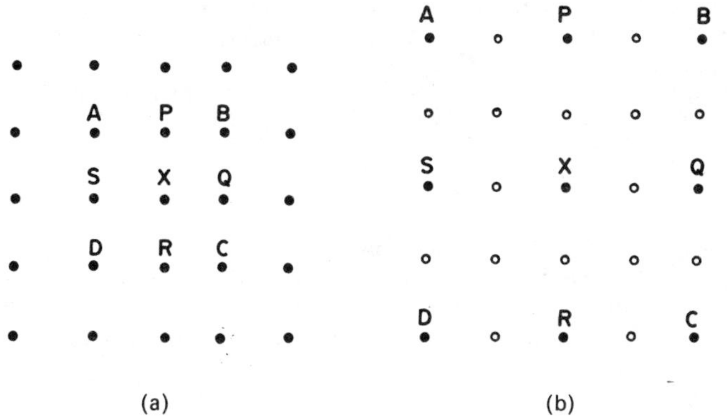

```
A      P      B
●   ○  ●   ○  ●

A P B
●   ●  ●  ●   ●

S X Q
●  ●  ●  ●   ●

D R C
●  ●  ●  ●   ●

●   ●  ●  ●   ●
```

(a)

```
A      P      B
●   ○  ●   ○  ●

○   ○  ○   ○  ○

S      X      Q
●   ○  ●   ○  ●

○   ○  ○   ○  ○

D      R      C
●   ○  ●   ○  ●
```

(b)

FIG. 4.12. Matter creation in the steady-state Universe: viewed from X, the galaxies A, B, C, D, P, Q, R, S are seen as in (a) at an earlier epoch. At a later epoch, shown in (b), they have moved away. New galaxies shown by open circles are created to fill up the gaps.

theoretical physicist, who likes to work from physical laws quantified by mathematical equations. These were lacking in the Bondi–Gold approach.

The creation of matter

Hoyle's approach to the steady-state Universe was aimed at answering the questions posed above in the language of a theoretical physicist. Here I shall summarize the basic ideas of his approach, without going into details.

If we take the steady-state theory as given by Bondi and Gold, we can ask the question, 'What is the estimated rate of creation of new matter?' The answer to this question depends on the mean density of matter in the Universe and the rate of expansion of the Universe (see Box 4.10). The estimate turns out to be $4 \cdot 5 \times 10^{-45}$ kg m^{-3} s^{-1}.

Box 4.10 The continuous creation of matter

The perfect cosmological principle tells us how to estimate the rate of creation of matter required to maintain the Universe in a steady-state.

Imagine a cube of space in the expanding Universe, with each side of length 1 m. If ρ is the density of matter in kilograms per cubic metre, the amount of matter in this cube is simply ρ. Now, by Hubble's law each side of the cube is expanding at the rate H, where H is Hubble's constant. So 1 s later the side of the cube will have grown approximately to $(1+H)$, and its volume will have grown to $(1+H)^3$. Since H measured in reciprocal seconds is very small, being given by

$$H \cong 1 \cdot 5 \times 10^{-18} \text{ s}^{-1}, \tag{1}$$

the volume increase is given approximately by

$$(1+H)^3 - 1 \cong 3H. \tag{2}$$

Now the PCP tells us that a second later the Universe should look the same as before. So its density must remain the same. The additional space created by expansion must therefore have the same density of matter ρ. Hence new matter has to be created in the one-second interval, with the mass

$$M = 3H\rho. \tag{3}$$

This is the rate of creation of matter. Observations suggest an estimated density of visible matter as not exceeding

$$\rho \cong 10^{-27} \text{ kg m}^{-3}, \tag{4}$$

so that from (1) and (3) we get

$$M \cong 4 \cdot 5 \times 10^{-45} \text{ kg m}^{-3} \text{ s}^{-1}. \tag{5}$$

This may have to be pushed up by a factor of between 10 and 100 to include any invisible matter present in the Universe (see p. 247).

In this calculation we have applied the PCP very rigorously. Bondi and Gold meant it to be applied in the statistical sense to large regions of space (measured in distances of several megaparsecs) and large intervals of time (thousands of millions of years). So the above rate is meant to be an average one, and is subject to fluctuations.

That is, for a kilogram of matter to appear in a cubical box with a side of length 1 m, we would have to wait 7 million million million million million million years. Although this rate of creation is extremely small, it must nevertheless be explained in the framework of physics and this is what Hoyle proceeded to do.

At first sight it would appear that new matter must be created out of some energy reservoir, if we are not to violate the law of conservation of matter and energy. However, a careful examination shows that this idea runs into difficulty. As we create matter, the reservoir will be steadily depleted. Moreover, as the Universe expands, this also tends to dilute the reservoir. Suppose we have initially 10 units of energy per cubic metre in the reservoir. If we create a unit of matter per cubic metre from this we will be left with 9 units per cubic metre of energy in the reservoir. Over and above that, the expansion of the Universe would further reduce this, say, to 8 units per cubic metre. If the process goes on and on, eventually nothing will be left.

To get 'round this difficulty Hoyle imagined a reservoir of *negative* energy. To see how this works let us go back to the last example, but with the difference that the reservoir now has −10 units of energy per cubic metre. If we create a unit of matter per cubic metre, we are left with −11 units of energy per cubic metre. The expansion reduces the *magnitude* of the reservoir so that it becomes −10 units per cubic metre again. So in this case the processes of creation and expansion work in the *opposite* direction and are held in balance in the steady-state.

Hoyle called this reservoir the C-field. It has the simplest physical laws imaginable—even simpler than those of the electromagnetic theory (see Box 4.11). Although Hoyle tried a number of quantitative formulations of this idea, he finally settled on the simplest one, suggested by Maurice Pryce. Technically, the C-field has no mass, no charge, and no spin. It comes into effect only at the time when particles are created. Whenever a particle with a certain energy is created, a C-field of equal but negative energy is also radiated. *The overall energy is therefore conserved.*

The concept of negative energy is unusual, but not new. Even the theory of gravitation allows the possibility of negative energy (see p. 141). Whether or not negative energy reservoirs can exist in the Universe is one of the more intriguing problems of theoretical physics and cosmology.

Hoyle's formulation therefore provided a theoretical description of matter creation *with mathematical equations*. The framework he used was that of Einstein's general relativity, with the modification that the C-field is also included (see Box 4.11). As with the Friedmann models and Einstein's equations, we can have various solutions of Hoyle's cosmological equations, of which the simplest proves to be the steady-state solution. This solution has an advantage over the big-bang models in that there is no singular point where the mathematical and the physical descriptions break down.

Apart from the simple steady-state solution to the cosmological equations we can obtain more complicated (and perhaps more realistic) solutions. If, for instance, the creation and expansion processes are not in exact balance, the Universe may fluctuate around the steady-state solution. (see Fig. 4.13). Also the creation of matter

†Box 4.11 The C-field

The C-field was introduced by Hoyle to account for the continuous creation of matter in the steady-state Universe, *without violating the law of conservation of matter and energy*. Below is given a comparison between the properties of the electromagnetic field and the C-field.

The electromagnetic field

1. Electromagnetic radiation originates when a charged particle accelerates. The radiation takes energy and momentum from the particle and moves away with the speed of light.

2. The electromagnetic field accelerates a charged particle, but does not directly affect the neutral particles.

3. The electromagnetic field has an *attractive* gravitational effect. Thus a region of space–time containing electromagnetic radiation will attract matter. This effect appears through Einstein's equations of gravitation.

4. The carriers of electromagnetic radiation, the photons, are particles of spin 1, no charge, and zero rest mass, travelling with the speed of light.

The C-field

C-field radiation originates when a particle is created. The radiation has energy and momentum equal and opposite to that of the created particle, and it moves away with the speed of light.

The C-field does not affect the motion of any existing particle. It only influences the particles at the time of creation, when it determines the particles' energy and momentum.

The C-field has a *repulsive* gravitational effect. A region of space–time containing C-field radiation will repel matter. This effect appears through Einstein's equations of gravitation.

The carriers of C-field radiation are particles of zero spin, no charge, and zero rest mass travelling with the speed of light.

may not be uniform all over the Universe but may be concentrated in pockets, say in the nuclei of galaxies.

The most fundamental question in cosmology is, 'Where did the matter we see around us originate in the first place?' This point has never been dealt with in big-bang cosmologies in which, at $t = 0$, there occurs a sudden and fantastic violation of the law of con-

servation of matter and energy. After $t = 0$ there is no such violation. By ignoring the primary creation event most cosmologists turn
a blind eye to the above question. The steady-state theory at least
brings this question within the framework of physics.

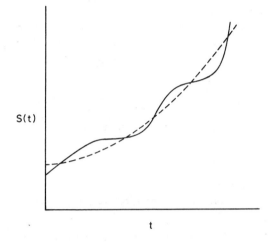

FIG. 4.13. The dotted curve represents strict steady state. The continuous curve indicates fluctuations.

Other cosmological models

Although the big-bang models, and to some extent the steady-state
model, have occupied the main cosmological stage so far, there are
many other models. Some of them have been based on Einstein's
theory of gravitation, but with less restrictive cosmological assumptions. Of these, we shall discuss the so-called 'spinning models'
briefly on p. 169.

Some theoreticians have constructed models outside the framework of Einstein's theory. For example, Einstein's theory requires
the gravitational constant G to be the same at all epochs, but there
are some observational and theoretical reasons for believing that this
may not be so. Models in which G changes with time have been
constructed by various theorists, for example, by Dirac as long ago as
1937 and later by Jordan, Brans and Dicke, and Hoyle and Narlikar.
Of these, the models by Brans and Dicke and Hoyle and Narlikar

make use of a proposition known as Mach's principle (see p. 165) as their starting points.

Dirac's cosmological model, however, is based on 'large-number coincidences'. Whatever their interpretation, the coincidences are remarkable in themselves and are worth mentioning.

Suppose we compare the electrostatic force F_E between an electron and a proton with their gravitational force F_G. Both F_E and F_G are forces of attraction obeying the inverse-square law of distance (r),

$$F_E = \frac{e^2}{r^2}, \quad F_G = \frac{Gm_em_p}{r^2}. \tag{4.9}$$

Here the proton has charge $+e$, and the electron has charge $-e$. Their respective masses are m_p and m_e. The ratio F_E/F_G can be easily calculated since e, m_p, m_e, and G are known. We get

$$\frac{F_E}{F_G} \simeq 10^{40}, \tag{4.10}$$

That is, the electrostatic force is about

10 000 000 000 000 000 000 000 000 000 000 000 000 000 (10^{40})

times larger than the gravitational force.

Consider next the ratio of the 'radius of the Universe' (R) to the 'radius of the electron' (r) (although both these are really misnomers). The characteristic dimension of the Universe is given by

$$R = \frac{c}{H} = cT, \tag{4.11}$$

where H is Hubble's constant, c is the velocity of light, and T is the reciprocal of H. R is the length over which significant changes in the structure of the Universe may be noticeable. As far as the electron is concerned, again it is not a rigid sphere with a definite radius, but the length

$$r = \frac{e^2}{m_ec^2} \tag{4.12}$$

represents a dimension characteristic of the electromagnetic and dynamical properties of the electron and is often called its radius. Clearly R is large and r small. What is the ratio R/r equal to? Again we find

$$\frac{R}{r} \simeq 10^{40}. \tag{4.13}$$

Is this an accidental coincidence that two apparently unrelated ratios turn out to be nearly equal, and so large at that? Or is there a connection between the two? This question has intrigued many theoretical physicists. It is further intriguing to note that the number 10^{40} is approximately equal to the square-root of the total number of particles in the observable Universe.

Dirac interpreted the result by saying that this coincidence reflects a connection between (4.10) and (4.13). He pointed out that (4.13) depends on the epoch of the Universe, since R depends on T. For (4.10) and (4.13) to be equal we would require (4.10) also to depend on the epoch. Here Dirac suggested that this could be achieved by letting G change with the epoch. If we equate (4.10) to (4.13) and assume that other quantities (which refer to atomic physics) remain constant, we get

$$G \simeq \frac{e^4}{m_e^2 m_p c^3 T} \propto \frac{1}{T}. \tag{4.14}$$

In a big-bang Universe the age of the Universe increases with T, and so (4.14) implies that G should decrease with the age of the Universe. The conclusions of Brans and Dicke and Hoyle and Narlikar, although arrived at differently, were qualitatively similar; that is, they predicted a decrease in G with time. In Chapter 5, we shall look at some evidence which might throw light on whether G changes with time.

Conclusion

A few decades ago cosmology was a subject with too many theories and too few observations. This is no longer the case. In Chapter 7 we will re-examine some of the theories that have been mentioned here in the light of the present observational data. But before coming to this confrontation between fact and theory in cosmology we will consider another approach to the cosmological problem. This approach consists of examining how the Universe as a whole influences the local environment and the physical laws which we study on the Earth. Although unconventional compared with the methods described hitherto, this approach is nevertheless equally important when studying the Universe. Amongst other things it tells us what types of cosmological model are consistent with local observations and what types are not.

5

What is Gravitation?

'Ask me another puzzle', said Alice.

'I bet you won't be able to do this one', said the Mad Hatter. 'I have two objects A and B; A is hot, B is cold. When I put them in contact with each other, A becomes hotter and B becomes colder. What are A and B?'

'This is impossible!' exclaimed Alice. 'I know for certain that heat passes from a a hot body to a cold body. So A should become colder and B should become hotter.'

'Think of stars.' The Dormouse gave a hint and promptly went back to sleep. But Alice had an objection: 'How can you put stars in contact with each other?'

'Trust you to raise silly objections!' said the Hatter. 'Even if I cannot do it in practice, I can surely perform a thought experiment?'

'But I still do not see how star A becomes hotter and star B becomes colder. This is against the laws of physics'. Alice was getting angry.

'Ha! Ha! You say so because you have not heard of gravitation', remarked the March Hare patronizingly.

It is clear that the law of gravitation plays an important part in the behaviour of astronomical systems. Even the overall structure of the Universe appears to be governed by this force. Yet when we examine the properties of gravitation we begin to discover its many puzzling aspects. Indeed, it would be no exaggeration to say that, although gravitation was the first of the fundamental laws of physics to be discovered, it continues to be the most mysterious one. In this chapter we are going to look at some of the strange properties of this force.

Laboratory techniques are not very effective for studying gravitational effects, because the effects are small. We saw in the last chapter how small the gravitational force is between an electron and a proton in a hydrogen atom when compared with the electric force between them. Thus unless we are dealing with huge masses, the force of gravitation is weak, and huge masses are to be found in the heavens rather than in the laboratory. It is to astronomy that we must turn for a possible breakthrough in our knowledge of the nature of gravitation.

Newtonian gravitation

About three centuries ago Isaac Newton discovered the simple, yet profound, law of gravitation. The law states that two objects with masses m_1 and m_2 situated a distance r apart, attract each other with a force given by

$$F = \frac{Gm_1m_2}{r^2}$$

where G is a constant, known as the gravitational constant.

We saw in the first chapter of this book how, by a simple calculation using the law of gravitation, we can estimate the mass of the Earth on which we live. And applications of the same simple law have enabled astronomers to understand the complicated motions of heavenly bodies—motions which puzzled such great minds as Hipparchus and Ptolemy. This explains why I used the adjective 'simple' as well as 'profound' to describe this law. By its successes the law of gravitation set the pattern for the subsequent development of physics. Even today physicists like their laws to be simple in statement, yet profound in implications. We now come to some strange consequences which follow from this simple law.

Negative energy

One of the consequences of Newton's laws of motion is the existence of a relation between work and energy. Suppose an external force acts on an object and moves it from one place to another. In this process the force is supposed to 'do some work'. The amount of work done by the force appears as the energy of motion (the 'kinetic energy', as it is called) of the object. When a football is kicked, the person kicking it does some work. This results in the motion of the ball and a gain in its kinetic energy. Now this gain in energy of the moving object is compensated by a loss of energy in the agency which is causing the working force. (In the above example the player kicking the ball will be a little tired as a result of the work done in kicking the ball.) This is an example of the well-known 'principle of conservation of energy'.

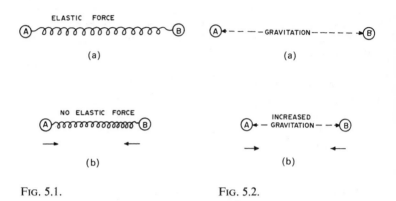

FIG. 5.1.   FIG. 5.2.

Let us now take another example (see Fig. 5.1). Here we have two balls A and B joined by a spring. Suppose we hold the two ends of the spring and stretch it. There is now a tendency in the spring to contract. If we let the ends go, the balls A and B will move towards each other as a result of the contraction of the spring. Thus there is a force of attraction between A and B which ceases when the spring acquires its natural length. At that stage A and B are moving towards each other. In terms of energy exchange this can be interpreted as follows. Initially when A and B were at rest and the spring extended, the system had zero kinetic energy and a certain amount of elastic potential energy (the latter arises whenever the spring is extended beyond its natural length). In the second stage this elastic energy has become zero, but A and B have kinetic energy, that is, the elastic energy of the spring is converted into the kinetic energy of the balls. There is, however, a limit to the kinetic energy A and B can have. This is set by the initial condition, that is, by the extension of the string beyond its natural length.

Let us now see what happens in Newtonian gravitation. Suppose we have two masses A and B held at rest at a certain distance r apart (see Fig. 5.2). When we release them A and B will move towards each other under the gravitational force of attraction. As in the previous example, they will acquire kinetic energy because they are moving. This comes at the expense of the gravitational potential energy, and so far this problem seems analogous to the previous example. But now we come to the difference. Unlike the second stage of the previous example, here we do not have a situation where the gravitational energy is zero and the kinetic energy a maximum. In fact, if they are point masses, *there is no limit to the kinetic energy* A *and* B *can acquire.* The closer they come the greater is the force of gravitational attraction between them and the greater are their velocities towards each other. Why does this happen? The answer is that the gravitational potential energy of A and B is *negative.* As A and B approach each other and acquire more and more kinetic energy, their gravitational energy reservoir is depleted and made *more negative.* And there is no limit to this process. In the elastic-spring case the potential energy is positive, and the process of energy transfer to A and B goes on until the potential energy reaches its minimum value of zero. The difference in the two cases is illustrated graphically in Fig. 5.3.

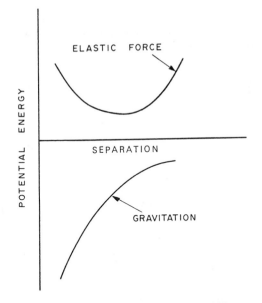

FIG. 5.3. The potential energy stored in the first system is always positive and has a minimum value. The potential energy in the second system is negative and steadily becomes more so as A and B come closer to each other.

All the forces in the Universe observed so far, except gravitation, have positive energy reservoirs like that of the elastic spring. If a system acted on by such forces gives in to them, the forces become *weaker*. In the elastic-spring example, initially the force between A and B is one of attraction. If A and B subject themselves to this force they come closer to each other. However, in this process the force of attraction is *reduced*. In the gravitational case, if the system gives in to the gravitational force, the force becomes *stronger*. In the example discussed above, if A and B get closer under the force of their gravitational attraction, this attraction force is *increased*.

If the laws of gravitation apply everywhere in the Universe, why don't we see this collapse happening everywhere? This question will be answered a little later on, but I will conclude our discussion of gravitational energy by considering a 'thought experiment' with apparently remarkable consequences.

It is our experience that if we place a hot body in contact with a cold body, heat passes from the former to the latter. This results in the hot body getting progressively colder and the cold body getting hotter, until the two have the same temperature. Consider now what

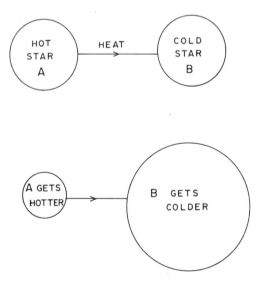

FIG. 5.4.

will happen in a thought experiment if we place a hot star A in contact with a cold star B (see Fig. 5.4). Heat will pass from A to B. However, we saw in Chapter 2 that the star A is in equilibrium under two forces: its self-gravitation and its internal pressures. The former tends to contract the star, and the latter prevents this. Now when heat passes from A, its internal pressures are reduced, and so it begins to contract. But this leads to compression of the gases inside A and a consequent rise in their temperature. In star B the reverse takes place. It gains heat from A, and therefore boosts up its internal pressures. This upsets the equilibrium inside B, and it expands. Expansion leads to the cooling of its gases. So, even if heat has passed from A to B, A has become hotter and B has become cooler. This paradoxical result arises again because of the negative energy of gravitation.

Energy transport

We have seen that radiation is a means of transporting electro-magnetic energy from one place to another. Does a similar situation exist in Newtonian gravitation? Can gravitational energy be trans-mitted from one point to another, say, with the speed of light? The answer is no. As mentioned in the previous chapter, Newtonian gravitation is an example of *instantaneous* action at a distance. Wave motion at the speed of light does not play any part in the Newtonian theory. Nevertheless there exists another means of transfer of energy between distant objects—it is called 'induction'. To explain this we shall follow the interesting experiment of Tweedledum and Tweedle-dee, conceived by two theoretical astronomers Hermann Bondi and W. H. McCrea.

To understand this experiment properly it is better first to under-stand how tidal forces operate in Newtonian gravitation. One of the well-known results of Newton's law is that a spherically symmetric distribution of matter exerts a gravitational force on an outside point mass as if the spherically symmetric object's entire mass is concen-trated at its centre. In Fig. 5.5 we consider three different cases of attraction by objects A,B,C. A is spherical. B is an oblate spheroid, (bun-shaped), and C is a prolate spheroid (egg-shaped), all of uniform composition. D is a point mass at the same large distance r from the centre of A, B, or C and lies on the plane of reflection symmetry. (If the object is reflected in such a plane its image coincides with itself. For a sphere any plane through the centre will do for this

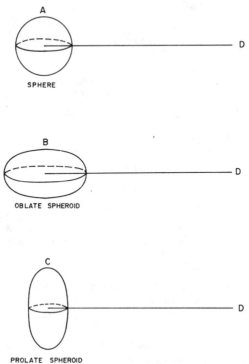

Fig. 5.5.

purpose; but in the case of the spheroids B, C there is a unique plane.) Now Newtonian gravitation tells us that if the mass of A is M and the mass of D is unity, the force exerted by A on D is

$$F_{AD} = \frac{GM}{r^2}$$

towards its centre. The forces exerted by B and C are different from this, even assuming that they have the same mass M. The force exerted by B is approximately of the form

$$F_{BD} = \frac{GM}{r^2} + \frac{a}{r^4},$$

where α is positive. Thus the force exerted by B on D is *greater* than that exerted by A on D. On the other hand, the force exerted by C on D is *less* than the force exerted by A on D. It has the form

$$F_{CD} = \frac{GM}{r^2} - \frac{\beta}{r^4},$$

where β is also positive. α and β depend on the actual shapes of B and C respectively. These extra forces are called 'tidal forces'. By Newton's third law of motion action and reaction are equal and opposite. Therefore the force exerted by D on A will be greater than its force on C but less than its force on B.

Suppose now that we have two identical spherical objects A, A' with their centres r apart. They will have a certain mutual force of attraction between them. Suppose that we change the shape of A into B. Because of the tidal effect this will increase the force between B and A'. But we can compensate this increase by changing the shape of A' to C. Of course the exact degree of prolateness required in C can be determined only after some complicated calculation. And on this fact hinges the experiment of Tweedledum and Tweedledee.

Tweedledum and Tweedledee are two identical spherical creatures; that is they are shaped like A. They have internal machinery which enables them to change their shape from that of A into an oblate form like B or a prolate form like C. Tweedledum and Tweedledee have been ordered to move in an elliptic orbit round each other under their mutual gravitation. However, they soon discover that tidal forces tend to distort their shapes, and as a result their orbits also get disturbed. To keep the orbits intact they must change their shapes suitably. Now, of the two, Tweedledum is clever but unscrupulous, while Tweedledee is honest but not very intelligent, and Tweedledum takes advantage of this to make Tweedledee sign an unfair agreement. Under this agreement, whenever Tweedledum takes up an oblate shape Tweedledee must take up the corresponding prolate shape which keeps their mutual attraction the same as if they were both spheres. Similarly, when Tweedledum becomes prolate, Tweedledee must become oblate. The exact degree of prolateness or oblateness required for this is to be calculated by Tweedledum, who will then inform Tweedledee. As we shall see Tweedledee pays heavily for this information.

Both Tweedledum and Tweedledee have electrical generators inside them which work in the following way. If a change of shape is effected against the dictates of the tidal force, the generators have to do work and they lose energy. If the change of shape is effected as required by the tidal force the generators do no work; instead they gain energy from gravitation. Now, the tidal force acts in such a way as to change a prolate body into an oblate one, and Tweedledum takes advantage of this fact.

During their orbit they are sometimes close to each other and sometimes away from each other (see Fig. 5.6). When they are close, the tidal force is very strong, and it tends to convert a spherical or a prolate body into an oblate one. At this stage Tweedledum becomes oblate and asks Tweedledee to become prolate. Naturally, Tweedledum's generators gain energy while Tweedledee's generators lose it. While they are far apart, Tweedledum becomes prolate and Tweedledee becomes oblate. However, in this case the tidal force is weak and Tweedledum does not have to spend much energy in changing his shape. Tweedledee gains some energy in this process, but it is only a small amount. When they come near each other again, the same process is repeated. Clearly, as they go round several times, Tweedledum gains considerable energy while Tweedledee keeps on losing it. And this is how the inductive transfer of energy can take place in Newtonian gravitation. The same tidal forces play an important part in the Earth–Moon system (see Box 5.1).

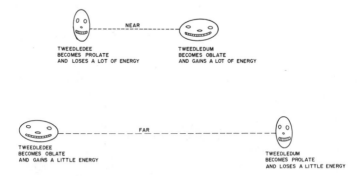

TWEEDLEDEE
BECOMES PROLATE
AND LOSES A LOT OF ENERGY

TWEEDLEDUM
BECOMES OBLATE
AND GAINS A LOT OF ENERGY

TWEEDLEDEE
BECOMES OBLATE
AND GAINS A LITTLE ENERGY

TWEEDLEDUM
BECOMES PROLATE
AND LOSES A LITTLE ENERGY

Fig. 5.6.   Energy transfer in Newtonian gravitation.

Box 5.1 The tidal phenomenon

The Moon and, to a lesser extent, the Sun exert gravitational forces on the Earth which give rise to tides. The effect is large when the two tides come together, that is, at new and full Moon times. Fig. 5.7 shows how the lunar tides operate.

If we imagine the surface of the Earth as covered with oceans at the ends A and A', the end A is pulled most strongly towards the Moon and the end A' least strongly. The main mass of the Earth which lies in between is pulled with a force intermediate to the two extremes. Thus the parts of the ocean at A and A' bulge outwards relative to the Earth.

Near land masses this results in a great inflow and a subsequent outflow of water from the sea. However, water moving over land experiences friction and loses energy. This energy comes from the Earth's rotational energy, with the result that the rate of rotation of the Earth about its axis slows down. This shows up in the lengthening of the day—by nearly a thousandth part of a second per century!

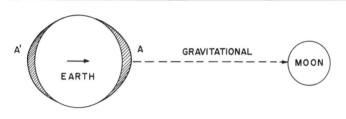

FIG. 5.7.

Gravitational collapse

We have seen how two objects attracting each other through gravitation move towards each other with ever-increasing velocity. If there is no force to stop them they will collide. Is the same thing true of a large number of particles attracting one another by Newton's inverse-square law of gravitation? The problem of many particles is, in general, not so simple; indeed we would require an electronic computer to sort out the motions of the various particles. However, in a special case the problem can be handled analytically.

Suppose we start with a very simple problem. We have a sphere of dust of uniform density with an initial radius R and mass M. By dust we mean material with no internal pressure (this is only an idealization; such a situation is not possible in practice). All the dust particles will attract one another, and the sphere will tend to contract. What is its rate of contraction? Suppose we measure the time from the instant the contraction sets in. Then at time t later, the sphere will have a radius r, which will be less than its starting radius R. To see how r changes with t, we note that a dust particle at the surface will have an acceleration towards the centre given by

$$a = \frac{GM}{r^2}.$$

This is a consequence of Newton's law of gravitation and his second law of motion. With the help of this equation the motion of the particle can be determined completely using the fact that at $t = 0$ (that is, initially), $r = R$ and the starting velocity is zero (see Box 5.2). The inward contraction rate at any time t is plotted against the

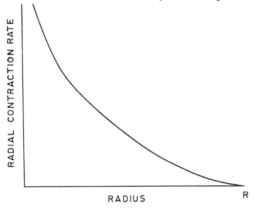

FIG. 5.8.   The collapse of a ball of dust.

---

†Box 5.2 Newtonian gravitational collapse

In the example discussed in the text, the equation of motion of the particle on the surface is given by

$$\frac{d^2r}{dt^2} = -\frac{GM}{r^2},$$

where the left-hand side denotes the outward acceleration. This can be integrated easily to

$$\left(\frac{dr}{dt}\right)^2 = 2GM\left(\frac{1}{r} - \frac{1}{R}\right),$$

where we remember that at $r = R$, dr/dt is zero.

The integral of the above equation, remembering that at $t = 0$, $r = R$, is given by

$$t = \left(\frac{R}{2GM}\right)^{\frac{1}{2}} \int_{r}^{R} \left(\frac{r_1}{R-r_1}\right)^{\frac{1}{2}} dr_1.$$

The time taken for r to become zero is obtained from the above formula by setting $r = 0$. This is the value t_0 given in the text.

---

radius in Fig. 5.8. We see that as r decreases to zero the contraction rate rapidly rises without limit. Clearly at this ever-increasing rate the sphere will shrink to a point in a very short time. This can be calculated, and is given by

$$t_0 = \frac{\pi}{2}\left(\frac{R^3}{2GM}\right)^{\frac{1}{2}}.$$

Suppose we have a sphere as large as the Sun in mass and radius. How long will it take to shrink to a point? All we have to do is substitute for R and M the appropriate values for the Sun. The answer comes out to be about half an hour; so that if the internal pressures of the Sun were suddenly removed it would shrink to a point in just over half an hour. It is, of course, the nuclear reactions in the Sun that provide sufficient internal pressure to hold it in equilibrium.

Now we return to the question which was posed earlier in this chapter. Why do we not see objects contracting rapidly under gravitation in everyday life? The answer is that *most* objects have sufficient internal pressures to withstand gravitational contraction. The word 'most' has been emphasized in the preceding sentence because exceptions cannot be ruled out, especially by the astronomer. From Newton's law it is clear that the gravitational force will be strongest in compact (R small), massive (M large) objects. Are there objects in the Universe whose self-gravitation is so strong that not even nuclear reactions can generate sufficient pressures to withstand it? With the discovery of QSOs (see p. 82) the existence of such objects became a distinct possibility. For this reason, there has of late been considerable interest in the subject of 'gravitational collapse', that is, the continued contraction (in a catastrophic manner) of objects under self-gravitation.

However, before we can go further with a discussion of this fascinating subject we must look at another theory of gravitation which is now preferred by most physicists for theoretical, if not practical, reasons.

The general theory of relativity

In 1905 Einstein proposed his 'special theory of relativity', which dramatically altered our previously held concepts of space, time, and motion. Although at first it encountered opposition and is still considered conceptually difficult, the theory has now gained wide acceptance among scientists. Indeed it is now part of 'what every theoretical physicist ought to know', and it forms the basic framework for describing all physical interactions.

About a decade later Einstein followed this up with another remarkable piece of work: the 'general theory of relativity'. Scientists may argue as to which theory represents the more radical departure from existing ideas, but it cannot be denied that both theories represent remarkable feats of scientific genius. In this chapter we are going to be concerned with the general theory of relativity. This is a theory of gravitation.

As a starting point in our attempt to understand what the general theory of relativity is about, we must first notice a remarkable property of gravitation, namely, its 'permanence'. If gravitational influence is present in a certain region, it cannot be switched off at any point, except momentarily. Take the example of the Earth's gravitational attraction. It is always there. We cannot construct a chamber inside which there is no gravity. A limited exception to this was suggested by Einstein in the now famous analogy of the falling lift. If a lift is falling freely (that is, without any controls on its motion) under the Earth's gravity, a man falling with it will feel weightless. While this remained a thought experiment in Einstein's lifetime, it has achieved reality in the present space-age. An astronaut going round the Earth is ideally under no forces except the Earth's gravity. So he is falling freely (see Fig. 5.9), although his initial velocity is such that he continues to move in a circular or elliptical orbit. While in orbit the astronaut feels weightless.

Now such examples of the cancellation of gravitational force apply only over a limited part of space and of time. If we take either of the two examples we see that weightlessness has only been achieved over a small region of space and over a small interval of time. Thus, in the astronaut's case, if we take a small region of space which includes

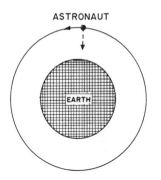

ASTRONAUT

EARTH

Fig. 5.9.

this orbit, gravity is absent over that region for the time during which the astronaut is crossing the region. Even here, a closer examination of Einstein's ideas shows that exact cancellation of gravity can be achieved only at a point of space and time, and there is still a small residual gravity present in our astronaut's space ship. Gravity cannot be destroyed permanently. This is in sharp contrast to the case of electric or magnetic fields, which can be destroyed in a permanent way. For example, the electric field in a region can be made zero by screening, that is, by surrounding its sources by earthed conducting surfaces.

Einstein interpreted this remarkable property of gravitation in the following way. He argued that its non-destructibility implies that it is, in a sense, permanent and all-pervading; and he related it to something else that has the same characteristics, namely, space and time. In what way can gravitation be related to space and time? Einstein achieved this through geometry, by arriving at a new synthesis of space, time, and matter.

The mathematicians of the nineteenth century had arrived at the conclusion that Euclid's geometry need not be the only possible geometry. By altering the basic axioms of Euclid's geometry, new geometries can be constructed which are entirely self-consistent. To give an example, take Euclid's parallel postulate (see Fig. 5.10). This states that, given a straight line l and a point p outside it, there is a unique straight line through p parallel to l. This appears to be a 'correct' result, yet it is just an assumption. It cannot be proved on

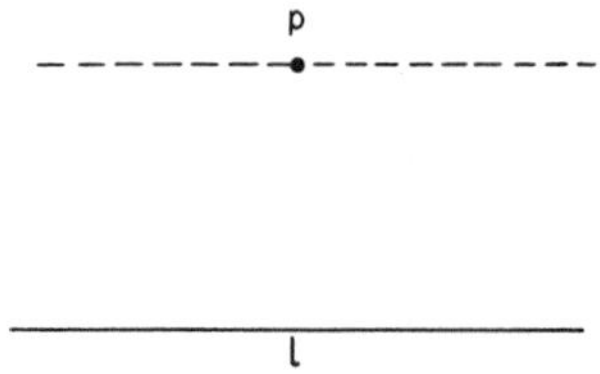

FIG. 5.10.

the basis of the rest of Euclid's postulates or axioms. The nineteenth-century mathematicians investigated whether something wrong shows up if this postulate is modified by saying either that *no* line through p can be drawn parallel to l or that *more than one* such line can be drawn. In either case they discovered no self-contradiction, and so the subject of non-Euclidean geometry was born. Theorems in a non-Euclidean geometry are different from those of Euclid. For example, the three angles of a triangle in such a geometry need not add up to 180° (see Box 5.3 for a practical example of one such non-Euclidean geometry).

Box 5.3 An example of non-Euclidean geometry

The surface of the Earth is not flat. Flat creatures crawling on the Earth's surface would not conclude that the geometry on its surface is Euclidean.

To see this, let us assume as a good approximation that the surface is spherical and that a flat creature begins a triangular journey starting from the North Pole, denoted by point A in Fig. 5.11. He starts southwards along the Greenwich meridian, reaching the equator at point B. He turns left and proceeds straight along the equator for a quarter of the way round the Earth, to point C. He then turns left along the meridian through C and reaches his starting point A. He discovers that he must turn left again in order to face in the same direction in which he started his journey. That is, he turned left *three* times, making a total turn of 270°, whereas a Euclidean triangle should have only 180° in its three angles! Except for making these turns, our flat creature has not deviated from a straight path; so he cannot be accused of not describing a proper triangle made of straight lines.

Einstein suggested that the effect of gravitation on the space–time

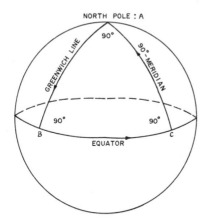

FIG. 5.11.

of any given region is to modify its geometry from Euclidean to non-Euclidean. Since gravitational influence is supposed to be exerted by all forms of matter and energy†, these form the 'sources' of the non-Euclidean geometry. Euclid's geometry and the *special* theory of relativity therefore hold only in an empty Universe. Since it is inconceivable to have the entire Universe emptied of all its contents, we can look upon the special theory as applying only in an approximate way, which explains the necessity for the nomenclature 'special' and 'general' as applied to the theories of relativity.

Einstein gave a set of equations to describe quantitatively how the non-Euclidean geometry arises through the presence of matter and energy. As an example consider the thought experiment where we draw a triangle ABC round the Sun (Fig. 5.12(a)). The 'straight' lines AB, BC, CA are the tracks of light rays. The three angles of this triangle will not add up to 180° but will differ from it by a fraction of the order

$$\theta = \frac{GM}{c^2R}.$$

† The equivalence of matter and energy had already been established by Einstein in his special theory of relativity.

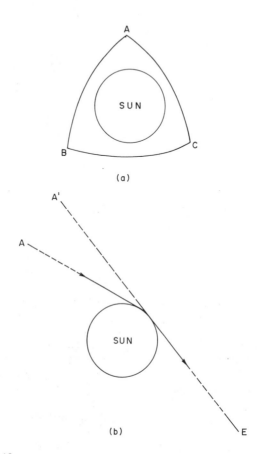

Fig. 5.12.

In this expression M and R are the mass and radius of the Sun, c is the velocity of light, and G is the same gravitational constant that appears in Newton's law. For the Sun this fraction is of the order one part in a million (10^{-6}). Although this is a thought experiment, a real experiment in a somewhat different form has been performed, and it seems to confirm the non-Euclidean nature of space as predicted by Einstein (see Box 5.4 for details).

Box 5.4 The bending of light

Suppose we observe a distant star A (see Fig. 5.12(b)) from the Earth E when the Sun S is close to the line of sight. According to the theory of general relativity, the light ray from A to E is bent by the gravitational field of the Sun, with the result that the star appears in a different direction A′.

So if the star is seen on two different occasions, say once when the Sun happens to be close to its line of sight and once when the Sun is well away from it, its direction should appear to change. There are many technical difficulties in making the first observation. For instance, for the star to be seen clearly with the Sun in the foreground the best occasion on which to see it is that of a total Solar eclipse, a rare phenomenon. Also we must eliminate carefully all bending effects due to other causes, for example, passage through the region close to the Sun. So the test is not all that conclusive.

The results so far show a deflection ranging from 1·43″ of arc to 2·7″, compared to the general relativistic value of 1·75″. If we adapt Newtonian gravitation and assume that photons, the light carriers, are attracted to the Sun like ordinary particles, we come up with *half* the relativistic value.

The fraction described above is a useful indicator of the departure expected from Euclid's geometry because of gravitating objects. It is usually very small, and this explains why Euclid's geometry is good for most practical calculations. In the same way its low value implies that quantitative predictions of Newton's and Einstein's gravitational theories will be substantially the same. That is why Newtonian gravitation has worked so well and even today is useful in computing the orbits of rockets and satellites. Because it is much simpler than Einstein's theory, Newtonian gravitation continues to be used in most calculations.

Astronomical tests like the one described in Box 5.4 have tended to give results in favour of Einstein, although the difference involved between Newton's and Einstein's theories is small. However, we will now investigate situations where the difference between the two theories becomes very marked.

Gravitational red-shift

According to Einstein, the effect of gravitation is felt not only in space but also in time. This becomes apparent in the discrepant rates at which clocks run in different regions. To be more specific let us consider two observers A and B (Fig. 5.13) in different regions of

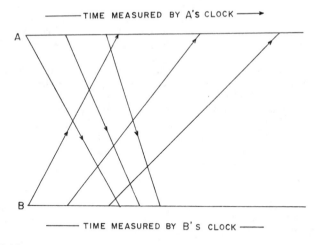

FIG. 5.13.

space. Near A there is a very weak gravitational influence—so that space geometry is almost Euclidean. Near B the gravitational influence is strong. Let us also assume that there is no change in the situation with time—that is, the situation is static. Suppose that A and B communicate with each other with light rays and that they decide to use atomic clocks in their respective regions as time-keeping devices. Now, from our everyday experience we would expect that if observer A sends signals every second by his clock, observer B will receive them every second, and vice versa. But here this is not what happens. To observer B, the signals from A appear to come at *shorter* intervals than a second, and conversely to observer A signals from B come at longer intervals than a second.

Such a situation can arise in astronomy. We may identify the region near A with the Earth and the region near B with the neighbourhood of a massive compact object. To observers on the Earth the clocks on the massive object will appear to go slower. In practice, of course, we do not see the clocks going fast or slow. Instead, we see changes in the frequencies of spectral lines, because these reflect the time changes in the atomic systems at the source. Thus in the above example, the frequency of light from the compact massive object will appear to be reduced, and the wavelength will appear correspondingly increased. This is nothing but the red-shift which we have encountered before. The fractional increase in wavelength produced in this way by an object of mass M and radius R at a place with negligible gravitational influence is given by

$$z = \frac{1}{\{1-(2GM/c^2R)\}^{\frac{1}{2}}} - 1.$$

Clearly this effect, known as the 'gravitational red-shift', is large if M is large and R small. For the companion of Sirius, which is a white dwarf star (see p. 40) the red-shift observed is about 60 parts in a million. More dramatic examples of the gravitational red-shift may occur in QSOs (see p. 236).

Gravitational collapse and black holes

Let us now reconsider the problem of gravitational collapse, looking at it this time from the standpoint of general relativity. Suppose we have a spherical body of mass M and radius R with no internal pressures—as in the Newtonian case. How will it contract?

The relativistic problem is more difficult, indeed it is impossible to solve explicitly in the general case. However, a simplified version does yield a solution. This is the case of a *uniformly dense* body. But even in this case further subtleties have to be taken into account.

In Fig. 5.14 we have a spherical body with uniform density undergoing contraction. As in the Newtonian case we consider a typical particle B on the surface of the body, and study its motion inwards. By way of comparison we have an external observer A located far away from the body, so that the observer is practically out of the body's gravitational influence. As the body contracts, the gravitational field in its neighbourhood increases, and the gravitational red-shift effect begins to assume significance. Suppose A and B are in communication with each other under the terms discussed on p. 157. The situation is different in one sense, however, because while A is at rest B is moving inwards away from A. This leads to dramatic consequences.

As in the static case on p. 158, B's clock appears to A to be slow. This effect arises from two causes: first from the gravitational red-shift and second from the Doppler effect (see p. 115), because B is receding from A. Both the gravitational red-shift and the Doppler red-shift are in the same direction. However, as seen by B, A's clock

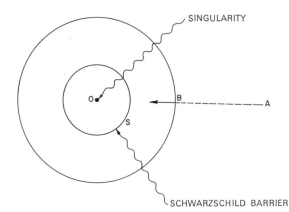

FIG. 5.14.

appears to be running either slow or fast—depending on whether the gravitational effect (which in this case contributes to the fast rate) is less or more important than the Doppler effect (which in this case contributes to the slow rate). It is clear, therefore, that the time scales of A and B are different. Let us study the situation both from the point of view of B and of A.

According to B there is a continuous collapse towards the state of infinite density. The interesting result is that with B's time scale the contraction rate of the body follows exactly the same rule as in the Newtonian case. Thus for a star of the mass and radius of the Sun (but of pressureless and uniformly dense matter) the collapse would occur in around half an hour. However, the similarity with the Newtonian case ends here. More drastic consequences are in store for B. As the matter in the collapsing object gets more and more dense, the geometrical properties of space–time become more and more peculiar (that is, non-Euclidean), until a state of infinite density is reached. At that stage all geometrical description breaks down—since the description involves mathematical operations with zero and infinity, operations which cannot be properly defined. This is the state of 'singularity'. It is very similar to the singularity of the big-bang Universe (see Chapter 4), except that there the Universe *explodes* while here the body *implodes*. (Indeed if we reverse the chronological sequence of events, the implosion becomes exactly like the big-bang explosion.)

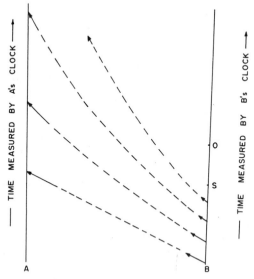

FIG. 5.15. Signals sent by B take a progressively longer time to reach A, as B approaches the Schwarzschild barrier S. No signals from B, after it crosses the barrier, reach A. B's own future terminates at the singularity represented by O.

What does A see during this period? Does he see B falling into the singularity? The answer is no, and the reason is as follows. B's signals arrive at A separated by intervals of more than one second. These intervals get progressively longer (see Fig. 5.15) as B falls further inwards—until a critical stage comes when B approaches a barrier known as the 'Schwarzschild barrier'.† When B reaches this barrier his signals no longer reach A, no matter how long A waits. All information about B at that time and after he crosses the Schwarzschild barrier is inaccessible to A. It must be emphasized that when B crosses this barrier he does not notice anything peculiar at all. By his clock everything goes on smoothly, until he reaches the singular state.

† First discovered in 1916 by K. Schwarzschild when he obtained the solution of Einstein's equations for a spherical gravitating object.

The Schwarzschild radius (that is, the radius of the spherical Schwarzschild barrier) for a body of mass M is given by a simple formula:

$$R_S = \frac{2GM}{c^2}.$$

For the Sun $R_S = 3$ km. This is much less than the Sun's actual 'radius', which is about 700 000 km. Only if the Sun shrinks from its present size to a radius of the order of 3 km will it become invisible to us.

Objects which are close to their Schwarzschild radius are almost invisible because light from them is highly red-shifted (red-shift of light implies a decrease in its frequency and hence its energy). Such objects are called 'black holes'. By definition, they cannot be 'seen'; but they can be detected through their gravitational influence. For instance, if the Sun were to become a black hole, it would cease to be visible but it would continue to attract the Earth. So the Earth would go round in an elliptic orbit with no apparent source.

The detection of black holes in the Universe forms one of the most intriguing searches of astronomy. It represents the ultimate negation of the conventional truth 'seeing is believing'. The X-ray source Cygnus X-1 is perhaps the most promising candidate for the location of a black hole so far known.

Whatever their observational status, black holes are keeping theoreticians well occupied. In recent years a number of investigations have been carried out regarding the nature of such objects, their possible observational properties, and their unusual geometrical properties.

Perhaps the most relevant question about black holes is 'what sort of an object can become a black hole?'. And here we encounter the difference between the gravitational theories of Newton and Einstein.

Suppose we consider an object a million times or more as massive as the Sun. How can we hold it in equilibrium? In the Sun, nuclear reactions generate internal pressures to withstand its self-gravitation. However, if we make an object more and more massive, nuclear pressures tend to rise in proportion to the mass, whereas the self-gravitational force rises in proportion to the *square* of the mass. So, for an object with a million solar masses, nuclear reactions are unable to provide the necessary pressures to balance the gravitational force. Such an object would therefore collapse and become a black hole, unless during the contraction it somehow got disrupted and broken up into smaller objects.

Can nothing whatsoever prevent the gravitational collapse of a massive object? In Newtonian theory we could conceive of some 'new' agency with strong enough pressures to halt the collapse. In Einstein's theory the situation is different. If we invent any such agency, its pressure must be accompanied by energy. This energy itself attracts and therefore helps the collapse. In the late 1960s work by theoreticians Roger Penrose and Stephen Hawking, has shown that, in general, unless we introduce new agencies with *negative energy*, collapse into a singularity is inevitable for most physical systems which have already contracted beyond a certain limit. Thus in the example discussed on p. 160, the collapse of B into a singularity cannot be halted once it has passed, say, the Schwarzschild barrier.

Are singularities desirable in a physical theory? Normally physicists and mathematicians tend to frown on such instances and regard them as indicators of imperfections in the theory. Accordingly it is possible to take the view that singularities in general relativity are undesirable and that we should look for 'better' theories. But there is the alternative view, formed probably because no better theories are in sight, that singularities tell us something unusual about the Universe and that, as such, they need not be objected to.

Gravitational radiation

What does the general theory of relativity say about gravitational radiation? The answer, unfortunately, is not as unambiguous as a physicist would like it to be. The reason for this can be explained qualitatively as follows.

In electromagnetic radiation we know that it is the electric and magnetic fields which propagate as waves with the speed of light (see Box 2.2). What propagates in gravitational radiation? We have already said that the gravitational effects in relativity are intimately related to the geometrical structure of space–time. So we expect the structural changes in space–time to propagate as 'gravitational waves'. In practice it is very difficult to single out any particular quantity which relates to such changes of space–time structure and which we can claim to be propagating as waves.

The difficulty lies partly in the coordinate description of space and time. Every physical observer is expected to use a coordinate system to describe the geometrical properties of space–time. For instance, he may need measuring rods and angle-measuring devices for spatial measurements and clocks for temporal measurements. Einstein's equations have the beautiful property that they have the same formal structure, whatever the coordinate frame of reference used. But this property gets in the way of deciding whether a particular solution does represent a gravitational wave. For example, in a wave propagation disturbances recur periodically. Even if we have such a periodicity in our solution, we cannot always be sure that this represents a genuine wave effect. The periodicity may appear as a result of the choice of a particular frame of reference. For example, a person being tossed up and down in a boat at sea may conclude that he is stationary and that his surroundings are going up and down, whereas in fact the effect is due entirely to the movement of his own frame of reference.

When gravitational fields are strong and the geometrical properties of space–time are very different from Euclid's, the problem of interpreting a disturbance as a gravitational wave becomes very difficult. But in the case of weak gravitational fields it is simpler to identify certain disturbances as gravitational waves. For example, it can be shown that two stars going round each other emit gravitational waves (see Fig. 5.16).

From an experimental point of view the situation is far from clear. Since 1968, Joe Weber, of the University of Maryland, has been detecting what he claims to be gravitational waves coming from the direction of the centre of the Galaxy. If he is right, the rate at which energy is radiated is estimated at between one to a thousand solar masses per year. This estimate is uncomfortably high from the point

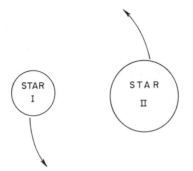

FIG. 5.16. Stars going round each other emit gravitational waves.

of view of galactic stability. Other observers who have since made independent efforts to detect this radiation have not found any positive result. At present this controversy has still to be resolved.

Although we have been mainly concerned in this chapter with the gravitational theories of Newton and Einstein, there have been many other theories of gravitation. Of these we shall now consider some which have been inspired by Mach's principle—a very remarkable concept which dates back to the last century.

Mach's principle

While formulating the laws of motion, Newton was faced with the problem of defining frames of reference. The motion of a body must be related to some specified frame of reference. For example, when we say a train is travelling at 50 miles per hour we mean that this is its speed relative to the surface of the Earth. But to a passenger in the train, the train does not appear to move; instead the surroundings on the Earth seem to be rushing past.

The problem of frames of reference first becomes important when we are talking about Newton's second law of motion. This states that the acceleration of a body (that is, the rate of change of its velocity) is proportional to the impressed force. Quantitatively we write

$$\text{force} = \text{mass} \times \text{acceleration}.$$

The mass of a body measures its inertia. The more inert the body the greater is the force required to change its motion. However, this change of motion must relate to some frame of reference. To make this point clear let us consider a simple example. Suppose we tie a stone to a string and whirl it round in a circle (see Fig. 5.17). How do we describe the motion of the stone by Newton's laws ? We can do it in two ways. First we can regard the stone as moving in a circle. Then it has an acceleration towards the centre of the circle, that is, along the string. This acceleration is caused by the tension in the string as required by Newton's second law of motion. The second picture is as viewed by an observer going round with the stone. For him there is no motion, and hence the total force must be zero. But he sees a tension in the string and concludes that the stone is being pulled towards the other end of the string. To reconcile this with the second law he therefore has to invent another force on the stone in the opposite direction. This force, which is equal in magnitude to the tension in

166   THE STRUCTURE OF THE UNIVERSE

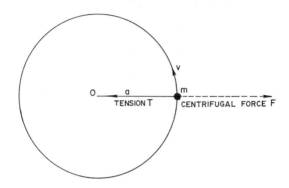

FIG. 5.17. The tension T is balanced by the centrifugal force F, in the rest frame of the mass m.

the string, is called the 'centrifugal force'. Note that this force has no apparent source; it has been invented in order to reconcile Newton's law of motion with the situation as observed from the stone.

Clearly there is something fundamentally different about the two modes of description. In one no additional force was needed, in the other it was necessary. Moreover, this additional force is proportional to the mass of the stone, and hence is called the 'inertial force'.

So we have two types of frame of reference. In one no imaginary inertial forces are needed to describe the motion. All existing forces account for the observed accelerations. Such a frame of reference is called an 'inertial frame'. In a 'non-inertial frame', inertial forces are required as described above. When Newton postulated his 'absolute' space (to which his laws of motion applied) he meant essentially an inertial frame. In his *Principia* he gives a graphic description of the difference between absolute and relative motions through his description of a bucket experiment:

The effects which distinguish absolute from relative motion are, the forces receding from the axis of circular motion. For there are no such forces in a circular motion purely relative, but in a true and absolute circular motion, they are greater or less, according to the quantity of the motion. If a vessel, hung by a long cord, is so often turned about that the cord is strongly twisted, then filled with water, and held at rest together with the water; thereupon, by the sudden action of another force, it is whirled about the contrary way, and while the cord is untwisting itself, the vessel continues for some time in this motion; the surface of the water will at first be plain, as before the vessel began to move; but after that, the vessel, by gradually communicating its motion to the water, will make it begin sensibly to revolve, and recede by little and little from the middle, and ascend to the sides of the vessel, forming itself into a concave figure (as I have experienced), and the swifter the motion becomes, the higher will the water rise, till at last, performing its revolutions in the same times with the vessel, it becomes relatively at rest in it.

Just as Newton's bucket experiment measures the rotation of water relative to absolute space, so we can design an experiment to measure the rotation of the Earth relative to absolute space. This experiment involves a Foucault pendulum, a pendulum which can oscillate in any vertical plane. The interesting result about its oscillation is that its plane of oscillation gradually changes its direction with time (see Fig. 5.18). Thus if it is set to oscillate in the north–south direction, a

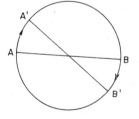

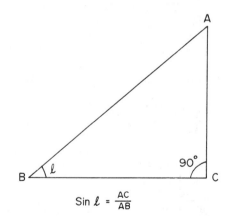

$$\text{Sin } \ell = \frac{AC}{AB}$$

Fig. 5.18. The pendulum-bob initially oscillates along AB. A few hours later it will be seen to oscillate along A′B′.

Fig. 5.19.

few hours later it will be found to oscillate in the east–west direction. The exact rate at which it changes direction is determined by the latitude of the place of observation. At a latitude l, the plane of oscillation goes completely round once in $1/\sin l$ days. (For a definition of $\sin l$ see Fig. 5.19). Thus at the North or South Pole the time taken is one day. At a latitude of 30° it is two days, and so on. This result has been obtained on the basis of Newton's laws of motion. The rotation of the pendulum plane takes place because the Earth itself is rotating relative to absolute space. A reference frame fixed on the Earth is therefore not an inertial frame. Additional inertial forces have to be introduced in order to describe the motion of the bob of the pendulum with respect to such a frame, and it is these forces which turn the pendulum's plane of oscillation.

The significance of all this is that someone in a closed room, and entirely unaware of the Earth's rotation, can discover and measure the rotation with the help of this experiment. By observing the time taken for the plane of the pendulum to go once round, and knowing the latitude of the place, he can conclude that the Earth completes one revolution about its axis in one day. This rate of revolution refers to Newton's absolute space.

There is, however, another way of measuring the Earth's rotation. This is by looking at distant stars. If we assume them to be fixed in the heavens, we can measure the rate at which the Earth is revolving about its axis. Naturally this rate of revolution is not relative to the absolute space of Newton but relative to the framework of distant stars. *The remarkable result is that the two rates of revolution turn out to be the same within observational errors.*

What does this mean? The immediate consequence of such a result is that the absolute space postulated by Newton is the same as the framework in which the distant stars (and galaxies) are non-rotating. But does the result imply something deeper?

The nineteenth-century philosopher and scientist Ernst Mach believed so. He maintained that motion can be described only relative to some concrete background—not relative to an abstract concept like absolute space. The distant stars provide such a background, and according to Mach (1893) their presence is absolutely essential to give any status to Newton's laws of motion. Earlier Bishop Berkeley (only a couple of decades after the publication of Newton's *Principia*) had also emphasized the important role of the stellar background. But Mach went further. He argued that the concept of inertia can be given a quantitative meaning only with the help of Newton's laws, which in turn require the background of distant stars. From this he concluded that inertia itself owes its origin to the background of distant stars. Remove the background and the body will cease to have any inertia! This reasoning is known as 'Mach's principle'.

Mach's principle, and its implication that inertia is not an intrinsic property of matter but is due to the background of distant stars, have received a mixed reception in the world of theoretical physics. Some physicists have taken the ideas with a grain of salt, arguing that they are all based on a mere coincidence of observations. Other physicists (including Einstein) were impressed by Mach's principle and tried to incorporate it into the rest of physics. The rest of this chapter is about some of these attempts.

Spinning Universe and general relativity

It was Einstein's hope that general relativity would turn out to incorporate Mach's principle. By establishing an intimate connection between space–time geometry and the physical properties of matter and energy, he did achieve what looked like a preliminary step towards Machian concepts. However, further investigations proved otherwise.

One explicit demonstration that general relativity does not incorporate Mach's principle came from Kurt Gödel in 1949. He constructed a model of the Universe out of Einstein's equations, in which the local inertial frame was not the same as the frame of non-rotating distant matter. On the contrary, distant galaxies in Gödel's Universe would appear to be spinning relative to the local inertial frame (in which Newton's laws hold). Gödel's model therefore denied the very observation on which Mach's principle is based.

Gödel's model used the λ-term in Einstein's equations. Also, it contained such peculiarities as closed time-like lines which permitted an observer to return to the same point in space–time after a while, and thus meet himself in the past! At first it was thought that the non-Machian character of the model probably has something to do with these characteristics. However, the later work of Oszvath and Schücking in the mid-1960s showed that even this is not the case. Spinning models *without* closed time-like lines were shown to be solutions of Einstein's equations *without* the λ-term. The existence of spinning Universes in Einstein's general relativity shows that Mach's principle is not fully incorporated in it.

The Brans–Dicke theory

In the early 1950s the Cambridge physicist Dennis Sciama suggested an interesting interpretation of Mach's principle. He argued that, when a non-inertial coordinate frame is used, the inertial forces arise because of gravitational forces exerted by distant matter. Imagine a body like the Earth which is being attracted by the Sun's gravitational field. In the frame of reference in which the Earth is at rest, we can argue that it is being acted on by two equal and opposite forces: (1) the Sun's gravitational force of attraction and (2) the force exerted by the rest of the Universe. The latter is expected to depend on the density of distant matter and its distance from the Earth. Starting with this idea Sciama deduced from general arguments the relation

$$\rho G T^2 \cong 1.$$

In this relation, ρ is the mean density of matter in the Universe and T is the time scale associated with the expansion of the Universe. If we use Hubble's constant H, we may write $T = 1/H$. Note that a similar relation followed in the various cosmological models which we discussed in Chapter 4, but here the approach is different.

In 1961 C. Brans and R. H. Dicke, from Princeton University, used a similar starting point, that is, defining the Machian influence of distant matter as a gravitational effect. They abandoned the concept—held sacred by Newton and Einstein—that the gravitational constant G is a fixed and fundamental quantity. Instead they sought to determine it in terms of the structure of the Universe as a whole. To do this quantitatively they introduced a new quantity called a ϕ-field. Brans and Dicke's ϕ-field is a *variable* physical quantity whose properties are determined by certain equations; its reciprocal acts as the gravitational constant. The fact that it is variable allows, in principle, for the strength of gravitation to vary from place to place and from time to time, depending on the overall structure of the Universe.

The Brans–Dicke theory has its own equations, resembling those of relativity theory but with the difference that ϕ is also present. The theory leads to consequences different from general relativity in many instances. Present observational tests seem to favour general relativity, but the issue still remains wide open. Like relativity, the Brans–Dicke theory also suggests a big-bang Universe, but with the difference that the value of G decreases with time.

The Hoyle–Narlikar theory

In 1964 Sir Fred Hoyle and I put forward another Machian theory of gravitation. In this we assumed, as a starting point, that the inertia of any piece of matter owes its origin to the rest of the matter in the Universe. This is the idea which had been suggested by Mach, but he did not give it a quantitative form. In our formulation such a form was given, and its consequences examined.

As in general relativity, we assumed that the geometry of space-time is non-Euclidean and that its character can be determined only with a full knowledge of all physical interactions in it. The equations of the Hoyle–Narlikar theory are obtained with the help of the so-called 'principle of least action', which is used as a starting point in most physical theories, including general relativity. The equations are much more complicated than Einstein's and have the following consequences.

1. When the number of particles in the Universe is large the equations can be approximated to look like those of general relativity—except near isolated particles. In these equations the gravitational constant G appears as a quantity determined by all the masses in the Universe. Thus, in general, the gravitational constant can vary as in Brans–Dicke theory, but in this approximation it remains constant and *positive*. The *sign* of the gravitational constant is important. In the theories of Newton and Einstein, G is tacitly assumed to be positive. In the Hoyle–Narlikar theory it is *deduced* to be positive.

2. The exceptional case mentioned above—that of an isolated particle—is interesting in the following way. By an isolated particle we mean a particle well separated from all others in the Universe. If we examine the motion of a test-particle† in a region in the neighbour-hood of such a particle, we find that it is first attracted by the isolated particle. But when the test-particle gets very close to the isolated particle *it is repelled*. Does this happen for two ordinary particles? This is an intriguing question, not yet answered. If gravitation changes sign and attraction changes to repulsion at very close range, this may explain why, for example, the massive galactic nuclei seem to be exploding rather than imploding. The problem is at present under investigation.

3. If we imagine the Universe to be empty, or to have just one particle in it, then the Hoyle–Narlikar theory predicts no interaction

---

† A test-particle is one which experiences the force of gravitation but does not exert it. It is an idealized concept.

at all. Thus a single particle will have no inertia. This is fully in accord with Mach's conjectures. In general relativity, on the other hand, even empty Universes have non-trivial gravitational properties, which do not conform with Machian concepts.

In 1971, Hoyle and I proposed another formulation of the model that throws new light on the singularity which occurs in the 'big-bang' Universes. In this formulation we allowed mass contributions to be both positive and negative from different regions of the Universe. This has many interesting consequences. First, the possibility exists of having regions of space where particles have zero mass. As we go away from these regions the magnitude of inertia increases. These regions of zero mass turn out to be closely related to the singular epoch $t = 0$ of a big-bang Universe. Indeed it turns out that a singularity appears in the big-bang Universe because of the insistence that the inertia of a particle should stay constant at all epochs. If we let inertia vary suitably with epoch the entire cosmological picture can be described in the flat, non-expanding space of special relativity. The galactic red-shifts arise in this picture not from the expansion of the Universe but because the masses of atomic particles were smaller in the past than they are now.

As I mentioned before, the other possible consequence of the Hoyle–Narlikar theory is that the gravitational constant should be slowly changing with epoch. It must be decreasing as we move away from the epoch when all particle masses were zero. We will now discuss how this variation in G may be detected.

Is the gravitational constant changing?

Both Newton's and Einstein's theories of gravitation assume the gravitational constant G to be constant in time. On p. 138 I described Dirac's cosmology, which requires G to decrease with the age of the Universe. The theories of Brans and Dicke and Hoyle and Narlikar make similar predictions. In all cases the rate of change of G as a fraction of G is of the order of Hubble's constant, that is,

$$\frac{\dot{G}}{G} = -aH.$$

Here $\dot{G}$ denotes the rate of increase of G with time. a is a constant which has a value equal to, or comparable to, unity.

Since H is estimated to be the reciprocal of nearly 20 000 million years, in the human lifetime this constant will not change by more than a few parts in 1000 million. So how can this be detected?

---

†Box 5.5 The measurement of time-variation of G

Sir Fred Hoyle has given the following argument to indicate how the rate of change of G with time can be measured.

If we consider the rate of change of the mean angular motion n_p of a planet round the Sun, we get from the relations similar to those given in the text

$$n_p \propto G^2, \text{ that is, } \frac{\dot{n}_p}{n_p} = \frac{2\dot{G}}{G}, \tag{1}$$

where a dot represents a rate of change. For the Moon, we get

$$\frac{\dot{n}_M}{n_M} = \frac{2\dot{G}}{G} + \left(\frac{\dot{n}}{n}\right)_{\text{tidal}}, \tag{2}$$

where the last term denotes the tidal contribution. So subtracting (1) from (2) we get

$$\frac{\dot{n}_M}{n_M} - \frac{\dot{n}_p}{n_p} = \left(\frac{\dot{n}}{n}\right)_{\text{tidal}}. \tag{3}$$

The advantage of subtraction is that it eliminates many systematic errors of observation, such as the use of astronomical time rather than the more accurate atomic time. (The former method of time-keeping is based on the Earth's orbit round the Sun and is subject to small variation. Unfortunately most of the old data are described with this time scale.) The disadvantage is that (3) now contains no information about $\dot{G}/G$!

To get this information we can use the ancient eclipse data. An eclipse gives a fairly accurate position of the Moon in relation to the Earth and the Sun, and we can go back to old records over several centuries for this. R. R. Newton has in this way estimated the value of $\dot{n}_M/n_M$. If this is the same as that given by (3) then clearly from (2) we get the result that $\dot{G}/G$ is zero. Newton's calculation shows

$$\frac{\dot{n}_M}{n_M} \simeq -0\cdot25 \text{ per aeon}, \tag{4}$$

where an aeon is a thousand million years.

On the other hand, the lunar and planetary data show that

$$\left(\frac{\dot{n}}{n}\right)_{\text{tidal}} \simeq -0\cdot13 \text{ per aeon}. \tag{5}$$

Thus there is a clear difference between (4) and (5), suggesting that

$$\frac{2\dot{G}}{G} \simeq -0\cdot12 \text{ per aeon.} \tag{6}$$

With $G \propto t^{-1}$, this implies $t \simeq 17\,000$ million years for the age of the Universe. However, (4) and (5) must be further checked before such a relation is substantiated. For $\alpha = 1$, as in the Hoyle–Narlikar theory, $G \propto t^{-\frac{1}{2}}$, giving half the above value for t. This is in agreement with the value of Hubble's constant as determined by Allan Sandage and his colleagues at the Hale Observatories in 1971.

One promising method seems to be the observation of the Moon's orbit round the Earth. Even if G changes with time the angular momentum of the Moon about the Earth (see Box 5.5) remains unchanged. Suppose this is denoted by h. Then the radius of the Moon's orbit and its mean angular motion are given by

$$r = \frac{h^2}{m^2MG}, \qquad n = \frac{m^3M^2}{h^3} G^2,$$

where m is the mass of the Moon and M is the mass of the Earth. Suppose we now allow G to decrease slowly. The above relations tell us that r will increase and n decrease with time. This also is obvious intuitively. As G decreases, the Earth's hold on the Moon should decrease, and the latter should move away from the former. Also, in order to conserve its angular momentum, its angular velocity should decrease in this process.

With the availability of atomic clocks this slow change in n is, in principle, measurable. The problem, however, is made more complicated because of tidal forces (see p.148). These also cause a slow variation in n with time. The way to get round this difficulty is shown in Box 5.5. Preliminary calculations along these lines do seem to show a variation of G as required by the above cosmological equation.

The variation of G affects the Earth in two ways. First, if G were larger in the past, the Earth must have been closer to the Sun. Also, the Sun itself must have been considerably brighter in the past than it is now, because, as a consequence of the equations of stellar structure (see Box 2.7), the luminosity of the Sun rises sharply with G. This means that the Earth received much more light from the Sun, say 3–4 times as much as it is getting now, when it first came into existence. Could a much hotter Earth support the beginning of life and biological evolution? This is an interesting question, and investigations so far do not seem to rule out the possibility that such a thing is possible and that there was a higher value of G in the past.

The second effect which varying G has is on the structure of the Earth itself. With a weakening gravitational effect the Earth will tend to expand, leading to a cracking of its outer crust. Is this how the continents were formed? The internal pressures of the fluid layers immediately below the outer crust of the Earth seem strong enough to move the broken crust about, and Sir Fred Hoyle has suggested that such pressures may well be responsible for the large-scale movements of the continents, that is, for continental drift.

It remains to be seen how strong a case can be made out for a decreasing G. No doubt, if it is well supported by observations, the entire superstructure of gravitational theory, built round Newton and Einstein, must collapse. Here, however, I have used the case of the varying G simply to illustrate how theories based on the large-scale structure of the Universe can lead to a drastic modification of well-established concepts of laboratory physics.

6

The Universe and the Arrow of Time

'The Sun would not radiate if it were alone in space and no other bodies could absorb its radiation . . . If for example I observed through my telescope yesterday evening that star which let us say is 100 light years away, then not only did I know that the light which it allowed to reach my eye was emitted 100 years ago, but also the star or individual atoms of it knew already 100 years ago that I, who then did not even exist, would view it yesterday evening at such and such a time . . '
H. Tetrode (*Z. Phys.* **10,** 317 (1922)).

We now come to a subject which has intrigued both scientists and philosophers as well as laymen. It has roused great controversies, very often due to misunderstandings but sometime arising because of genuine differences of opinion. The concept involved is that of the 'direction of time', and we will look at it here from the point of view of a physicist. The ways in which a physicist formulates his ideas of time are not always easy to understand. But the results are of such a unifying and all-embracing nature that anyone who makes the effort to understand cannot fail to be excited and stimulated by them.

The world of physics is four-dimensional; it has three dimensions of space and one of time. In this background of space–time, the laws of physics aim at giving a mathematical description of the behaviour of various systems existing in the Universe. The significant thing about all known laws of physics is that their mathematical formalism exhibits symmetries with regard to space and time. That is, with the limited exceptions mentioned in Box 6.1, the laws themselves do not

---

Box 6.1 Space and time inversions

Is there any method by which we can convey to an inhabitant of a remote planet what we mean by left and right or by clockwise and anticlockwise? If we try to convey this information through the laws of electromagnetism or gravitation, we will not succeed. These laws make no distinction between left and right. To put it another way, if we perform an experiment based on these laws and observe its consequences in a mirror, we would not see anything unusual.

However, not all basic laws of science show this property. In the mid-1950s it became clear that the law of weak interaction is not symmetric with respect to spatial reflection. The elementary particle, the neutrino, which interacts with other matter according to this law, shows the anti-symmetry in the following way. The neutrino is known to move with the speed of light. Besides this it has a spin which is *always* clockwise in the direction of its motion (see Fig. 6.1). Neutrinos

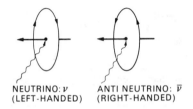

NEUTRINO: ν ANTI NEUTRINO: $\bar{\nu}$
(LEFT-HANDED) (RIGHT-HANDED)

FIG. 6.1. The neutrino spins in the clockwise sense in the direction of its motion. There are no neutrinos which spin the other way. There are, however, *anti*neutrinos which spin in an anti-clockwise sense in the direction of motion.

with opposite spin do not exist in the Universe. So we now tell our correspondent on the remote planet how to perform an experiment which releases neutrinos, and how to measure its spin. The sense in which he finds the neutrino to be spinning is to be called clockwise in the direction of its motion.

The laws of physics seem to be time-symmetric in the sense that if we change the sign of the time coordinate, the mathematical structure of the equations is unaltered. This symmetry is called T-invariance by the physicists, in contrast to P-invariance (P for parity) for the left–right symmetry.

Another closely related symmetry is that between particles and anti-particles. For an electron of mass m and charge $-e$ there exists in the Universe an anti-electron, or positron, of mass m and charge e. The two together annihilate each other to produce pure radiation energy. All electromagnetic interactions of the electron can be converted into those of the positron by changing the sign of the charge and modifying the equations suitably. This symmetry is called C-invariance (charge conjugation).

The three symmetries combined together go under the name 'CPT-invariance'. When applied in this form it means that for any physical interaction between particles there exists another between their anti-particles with left and right interchanged and with the sign of time reversed. It is believed that even though P-invariance is violated, as in the case of neutrinos, CPT-invariance still operates in the Universe. Thus corresponding to neutrinos there exist antineutrinos with spin in the anticlockwise sense.

In mid-1960s a violation of CP-invariance was discovered. This came in the form of decays of a class of particles called K-mesons. The theory based on strict invariance under CP (that is, change of particle to anti-particle and left to right) predicted certain limited modes of decay of these mesons. Experiments revealed, however, that these particles decayed in certain additional modes, thus implying that the rule of CP-invariance was being broken. If CPT as a whole is invariant, this experiment means that pure time-reversal invariance (T-invariance) is also violated in this case. In the late 1960s some physicists, like Y. Nee'man, have suggested this as a possible cause of the arrow of time.

make any distinction between left and right and between past and future. Yet in our everyday experience there is a clear distinction between these directions in space and time.

A closer examination shows that, while this distinction is made more or less by a convention with regard to directions in space, it is absolute with regard to time. If something happens to interchange the meanings of the terms 'left' and 'right' all over the world no significant change will be felt. In the new set-up there will be predominantly more human beings with hearts on the right. There will be more left-handed persons than right-handed ones. And so on. But while 'right-hearted', 'left-handed' persons are rare in our world, they do exist. In the case of past and future an interchange would produce a drastically new situation. In such a world we would see the evolution of life in a reverse direction. Besides, other seemingly impossible phenomena would take place. Broken cups would neatly join and mend themselves, and light bulbs would absorb instead of radiate light. Moreover, while we can freely move in space from left to right and vice versa, it has not been possible to invent a time machine which will enable us to move freely into the past and the future. This is why we usually talk of an 'arrow of time'—an arrow which always points from the past to the future and in which our experience always refers to what happened in the past.

Why should there be a time arrow?

Irreversible phenomena

The events which we see in everyday life can be classified into two categories: the 'reversible' ones and the 'irreversible' ones. A simple way of understanding the difference between the two types is as follows. Suppose we take a film of the event in question, and run the film backwards. Do we observe in real life the event shown on the film run in this way? If we do then the event is reversible; otherwise it is irreversible. Examples of reversible events are a ball bouncing on a floor in a perfectly elastic manner and a pendulum swinging in a frictionless medium. On the other hand, a cup falling on the floor and breaking in pieces is an irreversible event.

It is irreversible events which give rise to the notion of an arrow of time. We could very well answer the question, 'Why an arrow of time?' by stating 'because of irreversible events'. Indeed many physicists like to think of an extra law of physics which, in some form or other, postulates the existence of irreversible events. While this approach is suitable for *defining* the concept of a time arrow, it hardly takes us any further towards its understanding.

A more fruitful approach, in my opinion, consists of relating the different types of irreversible events found in the Universe in order to see what is their common characteristic and whether they can be traced to a common cause. In this chapter we will be concerned with such an approach, and with this view I will classify the various irreversible events into three different categories.

A large number of irreversible events occur in the subject which physicists call 'thermodynamics'. The common feature of such events is that they progress from 'order' to 'disorder'. The disorder in a system has been given a quantitative meaning in thermodynamics through the term 'entropy' (Box 6.2). Thus in an irreversible event

---

†Box 6.2 The entropy of a system

Entropy is a measure of the disorder in a physical system. It has been given the following mathematical definition in 'statistical mechanics' (a subject which attempts to explain the macroscopic behaviour of matter in terms of the statistical properties of its microscopic components).

Suppose the system is made of many different components or subsystems which may be numbered $1, 2, \ldots n, \ldots$. Each subsystem may be found in a number of possible states, with different probabilities. Thus one state may be more likely than another. Let w_n denote the probability of the nth subsystem being in the state in which it has been observed in the system. Then the entropy of the whole system is given by

$$S = -\sum_n w_n \ln w_n. \qquad (1)$$

In a changing system as the different components change their states the corresponding w_ns change, and so S also changes. While we could imagine S increasing or decreasing through such changes, in practice only an increase in S is seen. Thus the system evolves to a state of 'maximum entropy'. The rule for determining the probabilities is given by one of the basic assumptions of statistical mechanics.

Suppose we have a box divided into two compartments. In one compartment there is vacuum while in the other there is gas at a given pressure and temperature (see Fig. 6.2). If we make a hole in the

INITIAL STATE

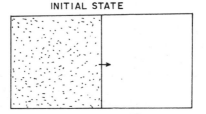

FINAL STATE

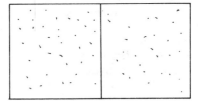

FIG. 6.2.

common wall of the compartments gas flows from the second compartment to the first until pressures in the two are equal. At each stage we can compute S for the entire system using (1). We will then find that the initial state has the lowest entropy and the final state the highest entropy. The same thing happens in the case of a cup falling on the ground and breaking.

the entropy of a system increases. For example, consider a cup dropping on the ground and breaking in pieces. Here the initial state is highly ordered, while the final state is one of disorder. Now the laws of dynamics (the most basic laws of physics) are time-symmetric and permit a situation in which all the different pieces are given just the right velocities so that they jump up and join together to form the cup. However, this sequence of events is never observed. To put this in general terms, the basic laws of physics do not lead us to think that the entropy of a physical system should increase with time, yet it always does.

A second type of irreversible event is to be found in the generation of electromagnetic radiation. We know that an alternating current (a.c.) circuit generates electromagnetic waves. These waves carry energy from the circuit outwards. The time-reversed version of this event, requiring energy to be drawn in by the circuit (see Fig. 6.3), is

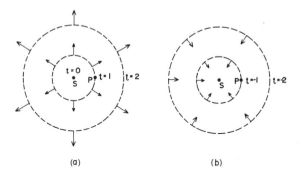

FIG. 6.3. (a) Retarded wave; (b) advanced wave.

never observed. Why? An appeal to the basic laws of physics does not help. The laws describing electromagnetic radiation—Maxwell's equations—are symmetric with respect to time. This means that just as they describe outgoing radiation they also allow the possibility of incoming radiation. Later we shall examine why incoming radiation is not found in the Universe. At present it is sufficient to note that its absence introduces an asymmetry with regard to time. An a.c. circuit loses energy by outward radiation, and if it receives no external electric supply it will eventually come to a halt. Suppose records are kept of the working of the circuit at various instants of time. If the records are all jumbled up, we can rearrange them in the proper chronological sequence, making use of the above time-asymmetry.

A third instance of time-asymmetry is provided by the Universe in the large. We saw in the preceding chapter that the Universe is expanding. Any two typical galaxies are found to be receding from each other. If we photograph their relative positions every thousand billion years (this is of course a thought experiment!) and mix up the photographs, we will be able to rearrange them in a chronological order making use of the fact that they are receding from each other. We can define the past and future of the Universe relative to the

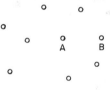

Fig. 6.4. The galaxies A and B were closer at an earlier epoch (a) and will be further apart at a later epoch (b).

present epoch by the density of any specified group of galaxies taking part in the expansion. In the past the galaxies were closer together than at present, while in the future they will be further apart (see Fig. 6.4).

Thus we have asymmetric events in thermodynamics, electro-dynamics, and cosmology. Each set of such events defines a time arrow. Is there any connection between these three arrows of time? Can we argue that the increase of entropy has any connection with the expansion of the Universe? Can we show any connection between the radiation phenomenon in electrodynamics and either the time-asymmetry of thermodynamics or that in cosmology?

Thermodynamics and cosmology

It is usual to regard thermodynamics as a branch of laboratory physics. Many thermodynamical discussions begin with such state-ments as 'suppose we isolate the system from its surroundings . . .' or 'let us enclose the system in a light-tight box'. While such state-ments simplify the mathematical discussion they obscure the impor-tant part played by the Universe as a whole. This important role is emphasized by the very simple calculation which was described on p. 9 under the title of the 'Olbers paradox', although there we left the paradox unresolved. Now it will be possible to give a satisfactory answer to the question, 'Why is the sky dark at night?' But before describing the correct answer let me mention some of the unsuccessful attempts at resolving the paradox, attempts made by Olbers himself and by others.

The Olbers paradox re-examined

Olbers made the following assumptions: (1) Space is infinite and obeys Euclid's geometry. (2) There is, on average, a uniform distri-bution of luminous objects, for example stars, in the Universe. (3) Each luminous object has the same luminosity (that is the same rate of output of luminous energy). (4) The Universe is infinitely old. (5) There is no absorbing matter in the intervening space to obscure the light from a distant luminous object. From these assumptions Olbers deduced that the sky would be infinitely bright if each source were a point source. If each source were a typical star like the Sun, and the more distant sources are occulted or hidden by the nearer ones, the sky brightness would match the surface brightness of the Sun. In either case, the conclusion does not agree with the observed darkness of the night sky.

One plausible way out of the paradox appeared at first sight to lie in the relaxation of the last assumption by the introduction of interstellar absorption. This, however, does not help. While, to begin with, the absorption diminishes the light from a distant source, it also heats up the absorbing material, which eventually begins to generate its own radiation.

The other alternative is to make the Universe finite in extent or finite in age. In the former case we only have a finite number of shells to deal with, and the answer is then finite. But this is a somewhat artificial device for resolving the paradox, especially since our present observations do not seem to indicate any bounds on the Universe. The finite-age idea does resolve the paradox, because of the finite velocity of light. If we assume the Universe was born a time T ago, and that the velocity of light is c, then the light reaching us *now* could not have come from *beyond* a distance cT. In this way we have an effective limit on the distance of the sources contributing to the brightness of the sky at present. While this is preferable to the finiteness of space, it has perhaps a similar touch of artificiality— instead of a finite extent, it is a finite age which now seems to save the situation.

The satisfactory resolution of the paradox came, however, with the discovery of the expansion of the Universe.

In an expanding Universe the major deviation from Olbers's original assumptions arises in the following way. The main sources of light are galaxies, and they are all receding from us. This leads to a diminution of light from a distant galaxy, over and above that arising from the inverse-square law, in two ways. First, there is a difference in the time scales which operate on the Earth and on the distant galaxy. It is this difference which results in the phenomenon of red-shift (see p. 121). Now this difference alters the rate at which light is received from the distant galaxy. The amount of light which is emitted by the galaxy in, say, one second, by its clock, is received here over a *longer* period. The factor by which the original one second is increased in the Earth clock is just the red-shift of the galaxy. Secondly, the red-shift implies a decrease in the frequency of light from the galaxy (which is carried by photons). This in turn implies a reduction in the energy which the light carries, by Planck's formula (see Box 3.4). So each photon, in its passage from its source to the Earth, loses energy. These two effects combine to reduce the contribution from distant galaxies in a significant manner, so that there is very little effect on the brightness or darkness of the sky due to the distant sources of light in the Universe. The complete answer to Olbers's question therefore is, 'The sky is dark at night because we are facing away from the Sun and *because the Universe is expanding*'.

Thermodynamics in an expanding Universe

The Olbers paradox can also be looked at from a thermodynamic point of view. Using all the original assumptions made by Olbers, we arrive at an infinite answer for point sources and a finite answer for extended sources, because the Universe as a whole tends to a thermodynamic equilibrium. This is the situation of maximum disorder (or maximum entropy)—when the entire space acquires the same temperature. In thermodynamic jargon, the stars form the 'source' and the rest of the Universe forms the 'sink'. Radiation from the stars passes heat continually from source to sink until the same temperature is established everywhere. This is possible in an infinitely old static Universe.

In an expanding Universe, on the other hand, the process of passage of heat from source to sink does not reach completion so easily, because expansion continually enhances the volume of the sink.

It is the same idea that led Bondi and Gold to *deduce* the expansion of the Universe from the perfect cosmological principle (PCP) (see p. 131). The PCP requires the Universe to look the same at all epochs. This permits three possibilities: (1) the Universe is static; (2) the Universe is ever contracting; or (3) the Universe is ever expanding. In cases (1) and (2) the situation discussed by Olbers would arise. Case (1) has already been discussed. Case (2) is worse from the point of view of brightness of the sky because in this case the radiation from distant objects will be 'blue-shifted', that is, increased in frequency and energy, which will lead to an infinite sky-brightness. Bondi and Gold noted that from observations of sky-brightness alone we can rule out cases (1) and (2), leaving only case (3). Since case (3) is consistent with the observed darkness of the sky, it, being the only possibility left, must be the correct one. Notice that this argument deduces the expansion of the Universe without any Hubble-type observation of the distant galaxies and without any mathematical theory of gravitation. It simply makes use of the PCP and the fact that the sky is dark at night, and thus illustrates the deductive power of the PCP.

Now we are in a position to see a possible connection between the arrows of time in thermodynamics and cosmology. If the cosmological arrow did not exist, that is, if the Universe were static, we would approach a thermodynamic equilibrium in which there would be complete disorder (maximum entropy). Such a Universe would have no physical activity which could define a thermodynamic arrow—it would have attained a state of 'heat death'. If, on the other hand, the Universe is expanding—that is, if a cosmological arrow of time exists—there is a departure from thermodynamic equilibrium. Thus the expansion of the Universe prevents a state of maximum entropy being reached by keeping the 'sink' partially unfilled. Hence a cosmological arrow of time must exist if a thermodynamic arrow of time is to exist.

The electromagnetic arrow of time

We now return to the electromagnetic arrow of time and to the question of why outgoing waves are present when an a.c. circuit is set up.

An alternating current consists of a to-and-fro motion of electric charges. One of the basic results to come out of Maxwell's equations is that an electric charge moving with constant speed in a straight line does not emit electromagnetic radiation. For an electric charge to emit radiation it must accelerate, that is, change its motion. This is precisely what happens in an a.c. circuit. So our problem really involves an investigation of what happens when an electric charge accelerates.

The two types of waves shown in Fig. 6.3 are called the 'retarded' and the 'advanced' waves respectively. These names arise for the following reason. Suppose we set up a source S (say an accelerated electric charge) at time $t = 0$ s. The outgoing wave begins its journey at this very time, and moves with the speed of light. So it will reach a point P a light-second away at a time $t = 1$ s. It will move on to further distances at *later* times. Hence the name 'retarded wave'. An advanced wave, on the other hand, moves outwards from the source at progressively *earlier* times. Thus an advanced wave set up at $t = 0$ s will be at P at $t = -1$ s. Since we are unaccustomed to counting time backwards, we choose to regard this wave as an *incoming* wave descending on the source S at $t = 0$ s. Thus we like to say that the wave was at P at $t = -1$ s and travelling *inwards* it reached S at $t = 0$ s. While this description is more convenient for our usual reckoning of time, it obscures the fact that the wave's source lies at S.

So the question of the electromagnetic arrow of time may be rephrased thus: 'Why are retarded waves present and advanced waves absent in the Universe?'. A simple but deceptive answer is often given to this question: 'because of causality'. 'Causality' is the name given to a principle which states that effects must *follow* rather than *precede* causes. It is believed that this principle operates in the Universe, and it is argued that since advanced waves violate this principle they cannot exist. However, this merely replaces our question by another, 'Why should the causality principle operate in nature?' Indeed the whole question of the arrow of time is about why certain chains of events (for example, cause and effect) occur in a fixed chronological order. Assuming the causality principle is tantamount to merely accepting this fact as a starting point.

An approach which goes deeper into the problem was initiated in 1945 by two American physicists John Wheeler and Richard Feynman. As a starting point they chose not the well-established description of electromagnetism as given by Maxwell but a re-vamped version of an older approach.

Action at a distance

To understand Wheeler and Feynman's approach it is instructive to look at the history of the subject of electromagnetism. It started as a set of two inverse-square laws—one for the attraction or repulsion of electric charges and the other for the attraction or repulsion of magnetic poles. Both laws, known as Coulomb's laws, were similar in structure to Newton's law of gravitation, and they seemed to give a fair description of the observed phenomena.

However, in the nineteenth century, discrepancies began to appear, especially when the motion of rapidly moving electric charges was involved. There were several phenomena which these laws could not explain. In 1845 the famous theoretician, Gauss, in a letter to Weber (another well-known physicist of that time) gave his conjecture as to the cause of these failures. He thought that the trouble lay in the fact that, according to these laws, electric and magnetic interactions propagated instantaneously. He felt that the interactions should really propagate with some finite velocity—he even made the prophetic suggestion that this could be the velocity of light. However, this idea was not followed up either by Gauss or by others.

In the 1860s, James Clerk Maxwell resolved the problem in an entirely different way. He abandoned the concept of 'action at a distance' between electric charges (or magnetic poles) and instead proposed the existence of a new entity—the 'electromagnetic field'. According to him two charges a and b interact in the following way (see Fig. 6.5). Both are surrounded by the electromagnetic field.

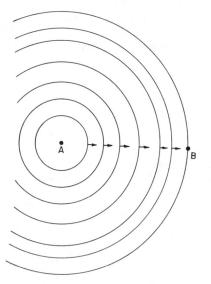

FIG. 6.5. The waves generated by the motion of charge a when at A propagate outwards and reach charge b at B. Then b feels the effect of a.

When a moves, it causes a disturbance in the field which propagates from a to b with the speed of light. This disturbance acts on b, and b begins to move. Thus the action of a on b is divided into three steps: the influence of a on the electromagnetic field, the propagation of the disturbance in the field, and the action of the electromagnetic field on b. These steps can be described mathematically (by a set of equations known as Maxwell–Lorentz equations).

The Maxwellian concept was very successful in describing diverse electromagnetic phenomena. Since success is the sole criterion by which scientific theories are judged, it is not surprising that Maxwell's approach completely replaced the earlier action-at-a-distance description of electromagnetism given by Coulomb.

Somewhat belatedly (in the present century), action at a distance was revived along the lines suggested by Gauss. Important work by Schwarzschild, Tetrode, and Fokker led to the formulation of *delayed* action at a distance. This is described in Fig. 6.6.

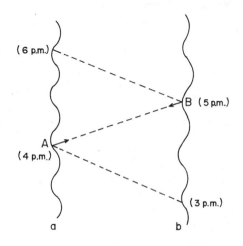

FIG. 6.6.

Here we can see tracks of two charged particles *a* and *b* in the space–time diagram. Let A be a typical point on the track of the charge *a*. From A draw a dotted line towards *b* to denote the track of a light ray. This intersects *b*'s track at a point B, say. What this diagram indicates is that the disturbance generated by the electric charge *a* at the space–time point A on its track is communicated to *b* not at once but at a later time denoted by B. Thus if *a* and *b* are at rest a light-hour apart and the disturbance from *a* leaves at 4 p.m., it will reach *b* at 5 p.m.

So far so good. But now a difficulty arises. Newton's third law of motion says that for every action there is an equal and opposite reaction. In the present picture it means that the action from A to B must be accompanied by a reaction from B to A, that is, *backward in time*. Hence if *a* acts on *b* via a *retarded effect*, *b* must react back on *a* via an *advanced effect*—and vice versa. Thus advanced and retarded effects are on an equal footing in this theory. This seemed to be so much against the experience in everyday life that the theory was again abandoned for a considerable period, until it was resurrected by Wheeler and Feynman, at the end of the Second World War.

The Wheeler-Feynman theory

To understand the Wheeler-Feynman approach, let us begin with a typical paradox that seems to arise when both advanced and retarded interactions are present in the Universe. Suppose two observers A and B, a light-hour apart, set up a system of communication with the following rules. The observer A sends out a signal to B at 4 p.m. if and only if he does not hear from B at 4 p.m. The observer B sends out a signal to A at 5 p.m. if and only if he hears from A at 5 p.m. Does this system work? It would work in a world of retarded interactions only—but not in a world of both advanced and retarded interactions. For if A sends a signal at 4 p.m. B will receive it both at 3 p.m. (via advanced waves) and at 5 p.m. (via retarded waves). Thereupon B will send a signal at 5 p.m. which will reach A at 4 p.m. and 6 p.m. But this violates the condition for A to send a signal at 4 p.m. If, on the other hand, A does not send a signal at 4 p.m., B does not send a signal at 5 p.m., and so A receives no signal from B at 4 p.m.—again violating the condition for A not sending a signal at 4 p.m.! Thus neither possibility is consistent.

Paradoxes like these are bound to arise if free communication between past and future is possible in the world. So a theory which permits situations like this must be abandoned.

Wheeler and Feynman, however, pointed out that so far as electromagnetism is concerned the above paradox does not represent the correct state of affairs. Suppose A and B in the above example are electric charges. In the actual Universe there are more than two charges. So when A and B set up their communication system they do not confine the electromagnetic signals to themselves alone. Take another charge C located *n* light-hours away from A. He gets A's signal at $(4+n)$ p.m. Because of this the charge C moves and generates its own influence which travels backward as well as forward in time. The former will arrive at A *n* hours *earlier* than when it was emitted, that is, $4+n-n = 4$ p.m. Note that this conclusion does not depend on n, that is, C's distance from A. In other words, the signal sent out by A receives an instantaneous response from C, no matter how far away C is located. We have already seen the vastness of the Universe. Unless we take into account the responses of *all* electric charges in it, we cannot determine the net outcome of the signal sent out by A. This important point was made by Wheeler and Feynman, and they then proceeded to calculate the net response of the Universe when an electric charge accelerates.

Wheeler and Feynman's work involves intricate mathematical details, but here we will try to get the gist of it in as simple terms as possible.

When an electric charge a is set in motion it generates, according to the action-at-a-distance theory, an equal mixture of retarded and advanced waves. Suppose we write this as the following combination:

$$F = \tfrac{1}{2} \text{ (retarded)} + \tfrac{1}{2} \text{ (advanced).}$$

Here (retarded) stands for the full disturbance generated if we accept the full retarded (outgoing wave) solution of Maxwell's equations. Thus we begin with complete symmetry between advanced and retarded waves. There is no prior commitment to the latter on the grounds of, say, the principle of causality.

The next step consists of setting a in motion and calculating the reaction from all other charges in the Universe. This may be called the response of the rest of the Universe to the motion of a. Using the same half-retarded-plus-half-advanced rule for all charges in the Universe, Wheeler and Feynman calculated the response of the rest of the Universe to the motion of a to be

$$R = \tfrac{1}{2} \text{ (retarded)} - \tfrac{1}{2} \text{ (advanced).}$$

When this response is added to the disturbance generated by a, the answer comes out to be

$$F + R = \text{ (retarded).}$$

In other words the response of the Universe neatly cancels the awkward part: the one which goes backwards in time. We shall refer to this as the 'correct response' from the Universe.

To understand why this happens it is instructive to follow a pictorial description (see Fig. 6.7) based on the argument given by Wheeler and Feynman. Here we see a sphere of large radius with a as centre; this sphere represents the distant charges in the Universe. Let b be a typical distant charge which has been excited by the disturbance of a. When b moves it generates advanced as well as

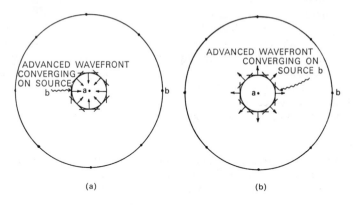

FIG. 6.7.

retarded waves. Let us consider the former as a spherical wave going towards *b*. As it approaches *a* it will be almost like a plane-wave surface moving along *ab*. The same will be true of advanced waves generated by all charges in this distant sphere. The plane-waves are all moving towards *a*, and their envelope generates a spherical wave converging on *a*. This state of affairs is represented in the first part of Fig. 6.7. The second part of the figure represents an instant later in time when all the plane waves have crossed *a* and are moving towards their sources. Now their envelope represents a spherical wave diverging from *a*. These converging and diverging parts are the advanced and retarded components of the response of the Universe. The 'rest of the Universe' in this context is called the 'absorber'. An essential condition for the response to be the correct one (that is, that given on p. 191) is that the absorber must be *perfect*—it must totally absorb all disturbance generated by *a*. If this condition is not satisfied, the response will fail to cancel the unwanted advanced part of the radiation. In an imperfectly absorbing Universe part or the whole of the disturbance generated by *a* escapes altogether.

A relation between the thermodynamic and electromagnetic arrows of time

Wheeler and Feynman performed their calculation in a static Universe. A static Universe has no cosmological arrow of time. If we change the sign of the time coordinate, the static Universe looks just the same as before. And this raises a paradoxical situation.

Suppose we do change the sign of the time coordinate in the Wheeler-Feynman argument. This leaves the Universe unchanged, but it interchanges the 'advanced' and 'retarded' waves. This means that we have a new response from the Universe which has exactly the opposite sign to that of the old response. This in turn results in the cancellation of the retarded half rather than the advanced half. The new solution obtained now must have the same validity as the old one obtained before. How do we resolve this ambiguity? Moreover, we are still far from achieving our objective, that is, answering why the Universe contains only retarded waves.

The ambiguity arises in the first place from the basic time symmetry of the picture. The action-at-a-distance picture is time-symmetric; the static Universe is time-symmetric. It will not be surprising therefore if we do encounter two solutions of equal status—one for retarded waves, the other for advanced waves.

Wheeler and Feynman recognized this, and they resolved the ambiguity by introducing the thermodynamic arrow of time into the picture. Thermodynamics, as we have seen, puts emphasis on the asymmetry of initial and final conditions. The former represents order, the latter disorder. Suppose we now apply this criterion to the source charge a and the absorber charges b, c, . . . In the first solution (that leading to retarded waves) the absorber charges are at rest *before* the disturbance from a hits them. *Afterwards* they begin to move and collide. This represents a transition from order to disorder, and hence is consistent with the thermodynamic arrow of time. In the second solution, on the other hand, the absorber particles are moving before the wave from a hits them and are at rest afterwards. *This represents a transition from disorder to order*, that is, in the reverse sense to that in thermodynamics. So whichever way we look at the situation, we find that the electrodynamic and thermodynamic arrows are pointing the same way. If we go from order to disorder, we are also using the retarded waves; if we go from disorder to order, we are using the advanced waves. The two solutions are therefore equivalent, and for convenience we may choose to work with the first one.

A relation between the cosmological and the electromagnetic arrows of time

In 1962 Jack Hogarth pointed out that the ambiguity which led

Wheeler and Feynman to think in terms of thermodynamics need not have arisen if they had worked in an expanding model of the Universe. In an expanding Universe there is a cosmological time arrow. If we change the sign of the time coordinate we also change the Universe from an expanding one to a contracting one—that is, to a new Universe. Suppose we find that the response in an expanding Universe cancels all advanced waves. It does not follow automatically from this that it will cancel all retarded waves in the other solution—simply because the transformation of time reversal (that is, change of sign of the time coordinate) is no longer permitted in an expanding Universe.

Hogarth himself, and later other authors including Hoyle, Narlikar and Roe, examined specific cosmological models to see which of them yield the correct response. The general rule for obtaining a correct response from a Universe model is as follows. We can divide the rest of the Universe with respect to the present position of the electric charge a as shown in Fig. 6.8. The two cones drawn with the charge at the apex represent 'light-cones' in the future and the past. These are

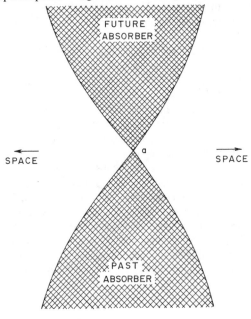

FIG. 6.8. The light-cones of the future and the past.

the tracks of light rays from a into the future and the past. The shaded regions in the interior of the cones represent the parts of the Universe accessible to physical disturbances arising from the motion of a. The cone and its interior in the future is the 'future absorber', while the cone and its interior in the past is the 'past absorber'. The rule for the correct response from a Universe model is that the future absorber is *perfect* and the past absorber is *imperfect*. If the reverse is true, the response is of opposite sign, that is, it leads to the cancellation of retarded solutions. If both absorbers are perfect, as in the static Universe, the result is ambiguous, and an appeal to another arrow of time, like the thermodynamic one, is necessary (see Box 6.3).

---

†Box 6.3 Perfect and imperfect absorbers

We have seen that a typical disturbance generated by an electric charge a in the action-at-a-distance theory is symmetric between retarded and advanced components. We write this as

$$F = \tfrac{1}{2} \text{ (retarded)} + \tfrac{1}{2} \text{ (advanced)}. \tag{1}$$

In a Universe with a perfect future and imperfect past absorber the response comes out as

$$R = \tfrac{1}{2} \text{ (retarded)} - \tfrac{1}{2} \text{ (advanced)}. \tag{2}$$

This response, added to F, produces the full retarded disturbance. How is this response generated? It is the full retarded disturbance from a which travels into the future absorber, disturbs its charges, and invokes the required response. So the argument is a circular one, although it is self-consistent.

To see the role of the absorbers, let us suppose a more general case where these are imperfect. This is expressed in the following way. The response produced by the full retarded disturbance from a is not R but fR, where f is a fraction less than 1. Similarly, the response produced by the past absorber to the full advanced disturbance from a is $-pR$, instead of $-R$, where p is a fraction less than 1. f and p therefore measure the efficiencies of absorbing power of future and past absorbers respectively.

Suppose such a Universe generates so much response that, when added to (1), it produces a total disturbance

$$T = \alpha \text{ (retarded)} + \beta \text{ (advanced)}, \tag{3}$$

where α, β are numbers which we shall presently determine.

As seen on p. 191 the full retarded field invokes a response R from a perfect future absorber. In (3) we have the full retarded field multiplied by the factor α, while the efficiency of the absorber is f. Therefore the retarded part of T generates a response $\alpha f R$ while for similar reasons the advanced part generates a response $-\beta pR$. For self-consistency, this response when added to F must give T, that is,

$$T = F + (\alpha f - \beta p)R. \qquad (4)$$

The coefficients of the advanced and the retarded components of the two sides of eqn (4) must match separately because the incoming waves cannot cancel the outgoing waves everywhere. Comparing coefficients we get

$$\alpha = \tfrac{1}{2} + \tfrac{1}{2}\,(\alpha f - \beta p), \quad \beta = \tfrac{1}{2} - \tfrac{1}{2}\,(\alpha f - \beta p). \qquad (5)$$

Solving these equations for α and β we get

$$\alpha = \frac{1-p}{2-f-p}, \; \beta = \frac{1-f}{2-f-p}, \; \alpha + \beta = 1 \qquad (6)$$

From (6) we see how the relative importance of advanced and retarded components of T is determined by f and p, which are the efficiencies of the future and past absorbers. We also see that if $f = 1$ but $p < 1$, we have $\alpha = 1$, $\beta = 0$, that is, the disturbance is fully retarded. This happens in the steady-state Universe. In an expanding big-bang Universe the opposite holds because $f < 1$, $p = 1$. We also see why ambiguity arises in the static or the expanding–contracting Universe, where $f = p = 1$. In this case both the equations of (5) merge into a single one

$$\alpha + \beta = 1.$$

So we cannot determine α and β except by bringing in additional considerations.

TABLE 6.1

Cosmological model	Past absorber	Future absorber	Nature of radiation
Static	Perfect	Perfect	Ambiguous
Ever-expanding Friedmann models	Perfect	Imperfect	Advanced
Friedmann model with expansion followed by contraction	Perfect	Perfect	Ambiguous
Steady-state	Imperfect	Perfect	Retarded

Table 6.1 summarizes the situation in the better-known cosmological models. It is clear that of those given in the table only the steady-state model gives the correct response. None of the big-bang models give the retarded solution in an unambiguous manner. To understand qualitatively the reason for this we have to take account of two effects present in an expanding cosmological model.

Take the case of the future absorber. All radiation emitted by a which travels in the future is red-shifted (see p. 121), that is, it gets diminished in frequency. So it is comparatively easy to absorb. However, in the ever-expanding big-bang models the density of matter in the future absorber decreases to zero so rapidly that there is not enough matter present in it to absorb even the low-frequency radiation. This results in the future absorber being imperfect. In the steady-state model, on the other hand, the density of matter is always the same and so there is enough matter in the future to absorb this radiation. Thus the future absorber in the steady-state model is perfect.

In the case of the past absorber the situation is different. The radiation going into the past is blue-shifted, that is, it has enormous frequency and energy. It cannot therefore be absorbed in the steady-state model, with its constant density of matter. In the big-bang models the density of matter in the past was very high—being infinite at the big-bang. Their past absorbers are therefore able to absorb even the highly energetic radiation coming from the future.

An ambiguous situation exists in the big-bang models which subsequently contract. Here infinite density in the past and the future ensures that both past and future absorbers are perfect.

Quantum physics

Our considerations so far have been on what we may call the 'classical' level of physics, and as such they refer only to a small part of the phenomena covered by the electromagnetic theory. In the first quarter of the present century it became clear to scientists that the real world is 'quantum' rather than 'classical' in its behaviour. That is, when we examine microscopic systems in nature, we find that they do not obey the laws of Newtonian dynamics. Instead of undergoing changes through continuous processes, these systems make transitions to other states in sudden jumps. To help us understand the difference between the classical and the quantum pictures of our world let us look at the hydrogen atom.

The hydrogen atom has a positively charged heavy particle, the 'proton', at its centre and a negatively charged but lighter particle, the 'electron', going round the proton. The attraction between the two may be described by Coulomb's law, and we would expect the electron to go round the proton in closed orbits, much in the same way as the planets go round the Sun. However, Maxwell's theory tells us that an electron moving in a closed orbit is always accelerated and hence must radiate energy. This energy comes from the total energy reservoir of the electron, consisting of its kinetic energy (that is, its energy of motion) and potential energy, arising from the attraction towards the centre of its orbit. The total energy decreases as the electron moves towards the centre (see Box 6.4). So the

---

†Box 6.4 The orbital energy of the atomic electron

Let us consider a classical electron of charge $-e$ in the hydrogen atom which has a proton of charge e at the nucleus. For simplicity suppose the electron goes round in a circular orbit of radius r with angular velocity ω. In a circular orbit its acceleration towards the centre, $r\omega^2$ arises from the force of attraction e^2/r^2 to the proton. By Newton's law of motion we have for an electron of mass m

$$m r \omega^2 = \frac{e^2}{r^2}. \qquad (1)$$

The kinetic energy T and the potential energy V of the electron are given by

$$T = \tfrac{1}{2} m r^2 \omega^2, \quad V = -\frac{e^2}{r}. \qquad (2)$$

From (1) and (2) the total energy of the electron is

$$H = T + V = -\frac{e^2}{2r}. \qquad (3)$$

Thus as its energy decreases by radiation, r decreases. So the classical electron is expected to spiral inwards and drop on the proton.

---

classical theory of Newton and Maxwell predicts that the electron will continually radiate energy and spiral inwards, eventually falling into the proton.

The hydrogen atom in practice behaves quite differently. It does not radiate energy, nor does its electron eventually fall into the central proton. Instead, the electron has various energy states (see Fig. 6.9). Each state corresponds to a specific orbit for the electron, in which *it does not radiate*. The electron can, however,

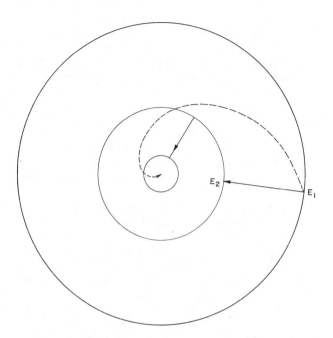

FIG. 6.9. The classical theory of Maxwell predicted that the electron in a hydrogen atom should continuously spiral inwards as shown by the dotted line, and fall to the central proton. In practice, the electron 'jumps' between discrete orbits. The innermost circle represents the 'ground' state. The electron does not go further inwards. This behaviour was explained by the quantum theory.

jump from one orbit to another. If it jumps from an (outer) orbit of higher energy E_1 to an (inner) orbit of lower energy E_2 it radiates a 'quantum' of radiation. The radiation has a frequency $\nu = (E_1 - E_2)/h$, where h is Planck's constant. There is, however, an orbit of minimum energy, the innermost orbit, in which the electron is said to be in the 'ground state'. Unlike the classical case, the electron in the ground state goes no further inwards, nor does it radiate energy. This is the quantum description of a hydrogen atom.

Now suppose such an atom is placed in a region filled with

electromagnetic radiation of frequency ν, arising from distant sources. Two things can happen. The electron can jump 'up' from an energy level E_2 to an energy level E_1 by absorbing energy from the ambient radiation, or it can jump 'down' from energy level E_1 to energy level E_2, releasing energy into the surrounding region. Such transitions are called 'induced' transitions. Note that this happens only if the radiation has a frequency corresponding to the energy difference between the two states of the electron.

The electron can also jump down and release energy even if no surrounding radiation is present. These transitions are known as 'spontaneous' transitions. Why do spontaneous transitions occur, and why are they always downwards? In Maxwell's field theory this question is answered by ascribing some unusual properties to a vacuum.

In classical physics the word 'vacuum' stands for a completely empty space, but not so in quantum physics. In quantum physics 'vacuum' means the state of lowest energy of the charged particles and the electromagnetic field. Fluctuations in the electromagnetic field are constantly taking place, leading to the creation and annihilation of electrons and their anti-particles, the positrons. This disturbed background leads to spontaneous transitions. (However, in the quantum version of the Wheeler–Feynman theory a vacuum is not required to play any such role; it is genuinely a vacuum. Instead the Universe is called upon to play an important part.)

In the quantum picture of an atom we can imagine the electron in a higher energy state being faced with three choices: (1) to stay where it is; (2) to jump down to a lower energy state; or (3) to jump up to an even higher energy state. Can we predict what course it will adopt? In classical physics every particle moves along a unique trajectory determined by Newton's laws of motion. In quantum physics there is no such unique path. Instead the electron can follow any of a number of geometrically possible paths, each path being given a certain probability. We cannot tell with certainty beforehand which path the particle will follow—we can only quote odds on various possibilities. It is this unpredictability that makes quantum physics vastly more rich in its effects than classical physics.

We now examine each of the electron's possible motions in the atom and ascribe probablities to them. This can be done by studying the consequences of each motion. Let us take the third possibility—that of jumping up. This involves acceleration accompanied by a gain in energy. In the classical physics problem of an a.c. circuit we saw that this is possible with an incoming-wave solution, that is, with the help of the advanced waves. However, in a Universe with the correct responses, like the steady-state Universe, no advanced waves exist. So this choice for an electron has zero probablity.

The second choice is the most interesting one. This involves acceleration and a loss of energy, that is, the use of retarded waves. This situation will therefore have a non-zero probability of occurring in a Universe with the correct response. This probability was calculated by Hoyle and myself in 1969, and turned out to be the same as that calculated for a jumping-down electron in electromagnetic field theory by using 'vacuum' fluctuations. However, the process in the Wheeler–Feynman approach is easier to visualize and more interesting to interpret. When an atomic electron spontaneously jumps down, it releases energy through radiation. This radiation causes upward transitions of the electrons in the future absorber. The excess energy gained by the absorber electrons is subsequently lost through dissipative processes like collisions. As in the classical version of the theory, the absorber produces a response at the source atom *instantaneously*. It is this response which causes the downward transition. This may look like a circular argument of the 'chicken-and-egg' variety. Did the electron jump first, giving rise to the response from the Universe? Or did the response come first, causing the downward jump? Indeed we cannot sort out the cause and the effect. Nor do we have to. We can simply compute the probability for the whole closed chain of events to take place. If this is less than unity (and it is) we conclude that there is a residual probability for the first choice (that is, for the electron to remain in its present orbit).

A new fact emerges from this argument. We have seen that for the future absorber to be effective its electrons must always jump up on receiving the radiation from the source. This is ensured only if they are mostly in their ground state. That is, the Universe must provide what is called a 'cold environment' in the future. Coldness here signifies low temperatures, that is, low random motions, and a state of low energy. It is easier to ensure this in an expanding Universe— since the expansion of any existing matter in the Universe produces a cooling effect, and here again we encounter the fact that the Universe must be able to act as a sink and to receive radiation. Thus we can once again see a connection between the three different arrows of time.

Why an arrow of time?

I hope I have made it clear that there is a strong connection between the three arrows of time: thermodynamic, cosmological, and electromagnetic. If we start with, say, the cosmological arrow of time, we have seen that, provided we live in a Universe with the correct response, we will also have the electromagnetic arrow pointing the right way. That is, the absence of advanced waves is not an arbitrary choice on the part of nature, but is dictated by cosmological considerations. Further, spontaneous transitions of the electrons in an atom occur not because of some magical properties of the electromagnetic field existing in a vacuum, but because of the expansion of the Universe, because of its ability to act as a perfect absorber in the future, and because of the cold environment which must exist in the remote future. This last aspect strongly emphasizes the connection between the thermodynamic arrow of time and the other two arrows. It is because the Universe expands, and radiation can leave its sources rather than descend on them, that we are able to see various thermodynamic phenomena. The expansion of the Universe prevents thermodynamic equilibrium being reached, and it is because a region is away from equilibrium that it exhibits a thermodynamic arrow.

The fact that the three time arrows all point in the same direction is a very significant result. Suppose we reverse the direction of the cosmological arrow. Then we should be in a contracting Universe. But will we as human beings 'see' it contracting? This is an intriguing question. If we take the viewpoint discussed in this chapter, we can make out a case for the electromagnetic arrow to be reversed also— that is, a case for advanced rather than retarded waves. The two arrows together would then reverse the thermodynamic arrow (although the case for this has not yet been made as strong as that for the electromagnetic arrow). This in turn may lead to the reversal of other time arrows, not discussed here at all, for example, the bio-logical and psychological time arrows. In that case we should be getting younger as the Universe contracts—which is the same thing as seeing the Universe expand as we grow older. What must be emphasized is that, from the observational point of view, the relative alignment of the various time arrows is more significant than the absolute direction of any particular time arrow.

To summarize, we may start with the time-asymmetry provided by our expanding Universe and try to understand the other instances of time-asymmetry found in the Universe. The ideas described in this chapter represent the attempts of many scientists to find unifying factors in the various manifestations of time-asymmetry. In particular, the statements 'the accelerated electric charge radiates energy because the Universe expands' and 'the atomic electron in a state of higher energy jumps spontaneously to a state of lower energy because the Universe expands' show the power behind this approach. To me, this connection between the smallest and largest systems in existence is one of the most attractive and exciting features to emerge from the application of physics to cosmology.

The Nature of the Universe: a Confrontation between Theories and Observations

Disciple: Revered Guru! Please explain to me the most salient feature of a good scientific theory.

Guru: I will do so with an example. Suppose I offer you a choice between two clocks: one is permanently stopped; the other gains a few minutes every day and has to be readjusted. Which one will you choose?

Disciple: I will choose the second one.

Guru: Are you sure? The second clock will never give you the exact time whereas the first one will give you the right time twice in twenty-four hours. So think again!

Disciple: Venerable One! What use is the first clock to me if I do not know when that right time is? I am satisfied with the second clock because I know that the time it tells is approximately correct within a few minutes.

Guru: Now you will appreciate what a good scientific theory must do. It must make predictions which can be tested by observations. Like the second clock, it may be imperfect in that its predictions are only approximately true. But it is always to be preferred to a theory which, like the first clock, has no predictive power.

In this chapter we shall take up the question of the origin and structure of the Universe from where we left it in Chapter 4. The theoretical models of the Universe described in Chapter 4 were based on certain assumptions suggested both by observations and by the criteria of simplicity. To what extent have these models been successful in describing the detailed properties of the Universe? Do they make testable predictions?

Although there are many different types of model of the Universe, the main thrust of the tests of validity has been directed against the steady-state model and the big-bang models. A good deal of cosmological controversy has been confined to the relative merits of the 'steady-state' and the 'big-bang'. While this controversy has contributed greatly to the development of cosmology as a branch of science, it gives a deceptively simple picture of the actual state of affairs, and this must be borne in mind when reading the next three sections.

The tests of cosmological models are of many different types, but they can be broadly classified into three categories.

Tests of models of the Universe by consistency

These tests are based on the following principle:'Our local environment must be consistent with the known physical laws and the structure of the Universe.' To apply this principle, a model of the Universe is assumed and certain laws of physics are taken as the starting point to work out a result directly applicable to the local environment. If the local environment fails to be consistent with that result either (1) the model of the Universe is defective or (2) the laws of physics assumed in the calculation are incorrect. We will look at two examples of this approach which have already been discussed in the previous chapter.

The Olbers paradox

The darkness of the night sky tells us a lot about what the Universe should *not* be like. If we start with a static and infinitely old model of the Universe which is uniform and infinite in extent, we end up with a contradiction with the observation of sky-brightness. Therefore such a model should be abandoned.

The arrow of time

If we take the Wheeler–Feynman theory (p. 196) as a starting point, then we can predict that, in the Friedmann models (the big-bang models which continually expand), the electromagnetic arrow of time is in the wrong direction. That is, accelerated electric charges do not *emit* outgoing (retarded) waves as seen in practice but rather they *receive* incoming (advanced) waves. This inconsistency may be removed either by abandoning such cosmological models or by abandoning the Wheeler–Feynman theory.

Tests of this type are perhaps the most clear-cut of them all, since the theoretical part is precise in its prediction and the observational part, being confined to the local region, is free from the ambiguities that accompany observations of the distant regions of the Universe. As in the second example, there may, however, be a dichotomy of conclusions: either to abandon the model or to abandon the physical theory. The choice, which may sometimes be subjective, should really depend on which alternative is more fruitful for physics or astronomy. Thus, if it should turn out that adoption of the Wheeler–Feynman approach leads to the solution of a number of outstanding problems in physics, then in the above test the big-bang models have to be abandoned. If, on the other hand, the Wheeler–Feynman approach does not prove fruitful, and the big-bang models are supported by other considerations, then such models should be retained and the Wheeler–Feynman theory abandoned.

Tests of models of the Universe based on the distant parts of the Universe

Observational tests using data from the distant parts of the Universe are called 'light-cone tests'. Since electromagnetic radiation is the common means of carrying such information, the velocity of light plays a central role in tests of this type. Any distant object in the Universe 'seen' through a telescope refers to an earlier epoch—to an epoch when the light which we receive *now* set out from the object. Therefore measurements of successively distant objects really refer to observations of the Universe on the past light-cone from the observer (see Fig. 7.1).

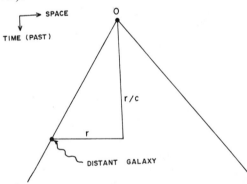

FIG. 7.1. A distant galaxy is seen as it was much earlier than the present epoch.

These tests are important because they reveal the non-Euclidean character of space–time geometry. The cosmological models, according to general relativity, are non-Euclidean in character, but this property becomes noticeable only over very large regions of space and long periods of time—just as the curvature of the Earth's surface becomes apparent only if we observe it over a large enough terrain. Unfortunately when we examine the distant parts of the Universe, ambiguities tend to creep into the observational data and its interpretation, and these tend to diminish the effectiveness of the tests considerably. Three tests of this kind are described below.

The red-shift–magnitude relation

This relationship is also known as the 'velocity–distance relation', and its use goes back to the very origin of modern cosmology. The present-day study of cosmology arose out of Hubble's observations of red-shifts from distant galaxies. When the red-shifts are plotted against the respective distances away from us of these galaxies, a straight line can be fitted to the plot. Moreover, if we relate the red-shifts to velocities of recession (as discussed on p. 117), the observed relation may be written in the form

$$v = HD,$$

where v is the velocity of recession, D the distance away of a typical galaxy, and H is Hubble's constant. Since the velocities are expressed in kilometres per second and the distances in megaparsecs (see Glossary for the definition of a parsec) Hubble's constant is often expressed in units of 'kilometre per second per megaparsec'. Hubble originally obtained a value of 530 km s^{-1} Mpc^{-1} which has since been drastically revised to 53 km s^{-1} Mpc^{-1}, that is, to a tenth of what it was. The dimensions of H are those of the reciprocal of time. So, $1/H$ is often expressed in time units, and its present value is about 20 000 million years—10 times Hubble's original estimate.

In the early 1930s Hubble's observations extended to velocities of the order of 1000 km s^{-1}, that is, to red-shifts not exceeding 0·004. Modern observations of galaxies go out to red-shifts of 0·63, representing more than a hundredfold increase over the distance covered by Hubble's observations. What do these new observations tell us about the Universe?

To examine this question, let us first consider the theoretical models. They all represent different non-Euclidean geometries. Now, over a small enough region, which in astronomical terms may mean up to a few megaparsecs (that is, up to the distances covered by Hubble), the different geometries do not differ significantly among themselves. To see this we go back to the analogy of the Earth's surface. Over distances of up to a few miles we can approximate geometry on the Earth's surface by Euclid's geometry on a flat surface. However, the 'flat Earth' description fails when we observe distances of a hundred miles or so, as, for example, any traveller on the high seas knows. In the same way, by looking out to greater and greater distances, where the differences between the various cosmological models become more and more appreciable, the astronomer hopes to pick out the one which matches the observations best.

These differences are reflected in the velocity–distance relation. Although all models predict the Hubble law as a linear (that is, a

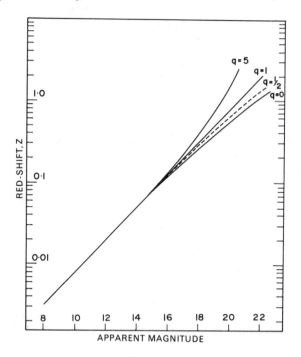

FIG. 7.2.   The theoretical red-shift–magnitude relation. The parameter q specifies a cosmological model. All curves agree for small values of z (and v).

straight line) law on a graph of v versus D, for small v and D, all the models deviate from linearity for distant galaxies in different ways. This is illustrated by the graph shown in Fig. 7.2. This graph does not plot velocity v against distance D but the red-shift z against the apparent magnitude. The reason for this is as follows. The observer does not measure either the velocity or the distance directly. Instead, he measures the red-shift and the apparent magnitude. Now we have seen that red-shift can be related to velocity, but the relation

$$v = cz$$

(obtained on p. 116) is only an approximate one, valid for small z. Indeed the formula is based on the concepts of Newtonian dynamics and Euclidean geometry—both of which are not applicable over the large distances covered in cosmology. It is the indiscriminate use of such ideas that invariably leads to paradoxical situations of the type discussed in Box 7.1. Rather than talk of velocity, it is much better to relate theoretical discussion directly to the red-shift.

Box 7.1 Super-light velocity?

Newtonian dynamics and the Doppler effect together lead to the following result for the red-shift z of a receding source of light

$$v = cz,$$

where v is the velocity of recession. The discovery of QSOs with z exceeding 3, often leads to the question, 'Are these objects moving faster than light?' This question arises from an incorrect use of the Doppler-shift formula.

Certainly, if Newtonian dynamics were valid for arbitrarily high velocities this formula would be correct, and there would be no contradiction involved in having objects moving faster than light. The ideas of Newtonian dynamics, however, are not valid except when v is very small compared to c. If a QSO has been thrown out of a nearby exploding object and happens to be moving radially away from us, we must use the theory of special relativity in calculating its motion. This gives the following formula, derived from special relativity,

$$v = c \, \frac{z^2 + 2z}{z^2 + 2z + 2}.$$

According to this formula v always stays less than c, and approaches that value only in the limit when z is infinitely large. For $z = 3$, $v = (\frac{15}{17})c$.

However, if the QSO red-shift is due to the expansion of the Universe it is wrong to use even the special-relativistic formula. This is because the geometry of space–time is now non-Euclidean and it is not possible to give a meaning to the velocity of a distant object in the same way as to the velocity of a nearby object. The cosmological red-shift is now given by the formula (see p. 121)

$$z = \frac{S(t_2)}{S(t_1)} - 1,$$

where $S(t_1)$ is the expansion factor at the time of emission and $S(t_2)$ is the expansion factor at the time of reception. For a given t_2 in a big-bang Universe, the light from the remote past is highly red-shifted because $S(t_1)$ is very small. Thus $z = 3$ implies that the linear separations between the galaxies at t_1 were a quarter of what they are at t_2.

The relationship between apparent magnitude and distance is even more dubious. You will remember (see p. 17) that the apparent magnitude is a measure of distance on a logarithmic scale. *If two galaxies are equally luminous intrinsically*, the more distant of the two will have the larger apparent magnitude. This follows from the inverse-square law of illumination (see Box 2.1). However, two uncertainties are introduced at this stage. First, the inverse-square law of illumination must be modified to take account of the non-Euclidean nature of the geometry. We have seen how this modification makes a significant difference to the Olbers paradox (p. 183). So the appropriate modification must be calculated for each cosmological model separately. The different curves shown in Fig. 7.2 are drawn on this basis. They are labelled by a parameter q, to which we shall return at a later stage.

The second complication lies in the observations rather than in the theory. Are we sure that all galaxies are equally luminous intrinsically? Even a survey of the local region shows substantial variations —by factors of 1–100 in the luminosities of galaxies. So in plotting their magnitudes we are liable to mistake a nearby but intrinsically faint galaxy for a distant but more luminous one. To avoid such mistakes as far as possible, Allan Sandage and his collaborators at the Hale Observatories have restricted their surveys to massive elliptical galaxies which happen to be the brightest members of their clusters (see Fig. 7.3 for an example). It is found that the luminosities of such galaxies do not show much variation. In Figs. 7.5 and 7.4 you can see a red-shift–magnitude plot using this subclass of galaxies in contrast to a plot of all galaxies. Evidently there is much less scatter in the former diagram.

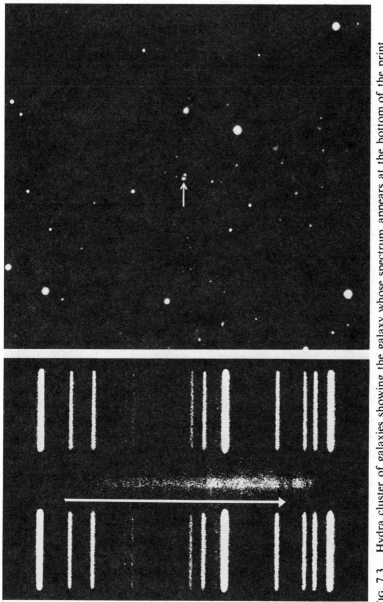

FIG. 7.3. Hydra cluster of galaxies showing the galaxy whose spectrum appears at the bottom of the print. Its red-shift is nearly 0·2. (Photograph from the Hale Observatories.)

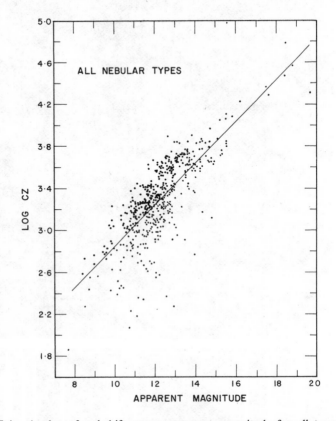

FIG. 7.4.   A plot of red-shift versus apparent magnitude for all types of galaxies. (After Humason, Mayall, and Sandage (1956). *Astronom. J.* **61**, 97.)

In Fig. 7.5 the plot is superposed on the theoretical plot of Fig. 7.2 to see which theoretical curve fits the data best. On the basis of this Sandage at first claimed that the curve labelled $q = +1$ fits the data best.

What does q signify? It is mathematically defined as

$$q = -\frac{S\ddot{S}}{\dot{S}^2},$$

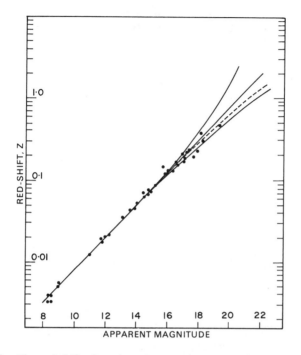

FIG. 7.5. The red-shift plotted against apparent magnitude for a subclass of galaxies chosen by Sandage. The theoretical curves
shown in Fig. 7.2 have been superposed on the plot.

which may be interpreted as follows. S is the expansion factor
described on p. 121. $\dot{S}$ denotes its rate of change and $\ddot{S}$ the rate of
change of the rate of change of S. So q essentially measures the rate of
slowing down of expansion. If q is positive it implies a slowing down
of the rate of expansion—if q is negative it implies an increasing
expansion rate.

So Sandage's observations seem to favour a model of the Universe
in which the rate of expansion is slowing down. Returning to the
Friedmann models, we can find a model of this type. It is illustrated
in Fig. 7.6. The model represents a big-bang Universe which expands
and then contracts. At present we are in the expanding phase. The
big-bang occurred, according to this model, about 11 500 million

years ago, and the total life of the Universe from explosion to implosion is about 125 000 million years. The Universe as a whole is finite but unbounded, like the Einstein Universe.

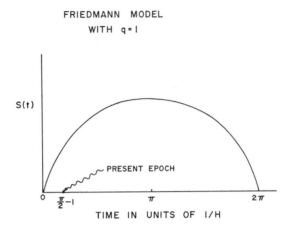

Fig. 7.6. A Friedmann model consistent with the rate of expansion slowing down (as suggested by observations by Sandage).

What are the underlying uncertainties behind this conclusion? First we note that the theoretical curves really begin to deviate at red-shifts in excess of $z = 0\cdot2$. So if several points were available on this graph beyond $z = 0\cdot2$, we could attach a certain measure of confidence to this conclusion. Unfortunately the number of such points is only three, with $z = 0\cdot29$, $0\cdot36$ and $0\cdot46$. The last of these represents a galaxy associated with the radio source 3C-295. This was discovered in 1960 by Minkowski, and it was then hoped that many more such cases would be found with the help of radio-source identifications. Although many radio-source galaxy identifications have since taken place, there have been none with a high enough red-shift to provide a better statistics.† So the Sandage estimate $q = 1$ cannot be taken with a great deal of confidence. Values $q = 0$ or even $q < 0$ are not inconsistent with the present data on galaxies. The

† Recently H. Spinrad has identified the radio source 3C-123 with a galaxy of red-shift $0\cdot637$.

steady-state model has $q = -1$, and even this case cannot be ruled out on the strength of the present data.

Another possible complication could arise from a significant change in the luminosity of a galaxy with red-shift. B. Tinsley, from the University of Texas, and her colleagues have made a case for a decrease in luminosity of a galaxy with age. This has the effect of lowering the earlier estimate $q = 1$ of Sandage. Many astronomers, including Sandage, now believe that q may be very close to zero.

For a clear-cut result we require a class of objects with not a great deal of variation in luminosity but with red-shift values far in excess of those found for galaxies. When quasi-stellar objects were first discovered, some with red-shifts close to $z = 2$, it was hoped that they would settle the cosmological issue once for all. By the present day nearly 500 QSOs are known, with a maximum red-shift amongst them of $z = 3.53$ (for the source OQ 172). Have QSOs settled the problem as expected?

Unfortunately QSOs have made the issue more complicated than before. The Hubble diagram for QSOs is shown in Fig. 7.7. It is a perfect scatter diagram, with no hint even of a connection between red-shifts and apparent magnitudes let alone an indication of the value of q. If we want to reconcile this at all with Hubble's law we will have to argue that there is an enormous scatter in the intrinsic luminosities of QSOs, and this explanation finds favour with many astronomers. But if this interpretation is accepted, QSOs cannot be relied upon to pick out the best cosmological model. To put it another way, imagine an inversion in the chronological order in which the red-shifts of QSOs and galaxies were discovered. Suppose in the late 1920s Hubble had come across QSOs and their big red-shifts. Would he, on the basis of Fig. 7.7, have concluded that they satisfy a velocity–distance relation? Indeed no unprejudiced observer, on the basis of this data, could have come to such a conclusion. It is because such a relation was found for galaxies in the first place that the astronomer is tempted to suppose that it applies to QSOs also, but it is quite possible to imagine the QSOs red-shifts to be of non-cosmological origin. We shall return to this controversial point later in this chapter.

To summarize, a somewhat frustrating situation exists with regard to this test. Theoretically, the different models predict really different results in the high-red-shift range. However, in this range, where the evidence from QSOs might have been decisive, a complete anarchy exists. In the low-red-shift range the galaxies do obey a systematic Hubble law, but here the observations cannot really decide which cosmological model is best suited, because most models of interest do not predict significantly different results.

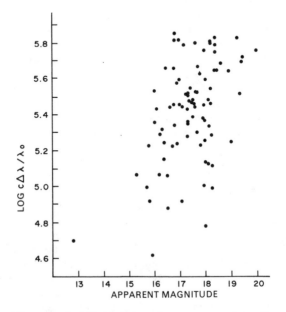

FIG. 7.7.   Plot of red-shift ($\Delta\lambda/\lambda_0$) against apparent (visual) magnitude for QSOs with red-shifts known in February 1967. (After Burbidge and Burbidge, *Quasi-stellar objects*, W. H. Freeman, San Francisco.)

The counting of radio sources

Besides the red-shift–magnitude relation, Hubble also undertook another cosmological test. This involved counts of galaxies up to different distances. The basic idea of the test is simple. If we suppose that (1) we live in a Euclidean Universe, (2) there is a uniform distribution of galaxies in the Universe, and (3) the galaxies are of the same luminosity, then we can perform the following simple calculation.

Let there be n galaxies per unit volume, and let each one have a luminosity L. Then the number of galaxies in a spherical region of radius R around us will be

$$N = \frac{4\pi}{3} R^3 n.$$

The faintest of these as seen from here will be those at distance R, and the rate at which light is received from any one of them over a unit area here will be

$$S = \frac{L}{4\pi R^2}.$$

From these we get the relation

$$N^2 S^3 = \frac{1}{36\pi} n^2 L^3 = \text{constant.}$$

That is, if we measure the number N of galaxies brighter than S the product $N^2 S^3$ should stay constant for varying levels of brightness.

If the product $N^2 S^3$ does not appear to be constant some or all of the assumptions (1)–(3) above must be wrong. The first two of these have a cosmological bearing, because they relate to the state of the Universe in the past.

Hubble's attempt in this direction did not succeed largely because the number of galaxies is fantastically large. In order to test any possible departure from the Euclidean nature of geometry, observations must cover large regions. It is estimated that some hundred million galaxies may have to be counted to get any significant information of this type! The test is feasible if it can be applied to a less numerous class of objects.

Radio-astronomy provides such a class. When the extragalactic nature of most radio sources became established, Sir Martin Ryle and his collaborators at Cambridge, England, set out to perform the above test for radio sources. For radio sources the quantity S is called the 'flux density', and it limits the radiation being measured to a narrow band of frequencies. The radio-astronomer plots $\log N$ against $\log S$ to test the constancy of $N^2 S^3$. If $N^2 S^3$ is constant the plot should represent a straight line with a slope of -1.5, as shown in Fig. 7.8. In the same figure is shown the line obtained by Ryle and his group from their survey. This line has a slope of -1.8, a significant departure from the -1.5, predicted on the basis of the assumptions (1)–(3). Assuming for the moment that (1) and (3) are still valid, the

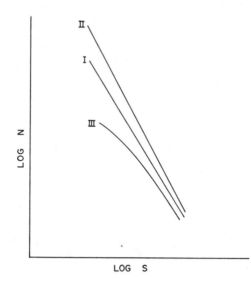

FIG. 7.8. I: Euclidean, II: Observed, III: Steady-state.

implication of this result is that the number n increases as we look at further regions, that is, as we look at regions further back in time (because radio waves take some time to travel from the source to the receiver).

Alternatively, we could argue in the following way. Using one of the theoretical cosmological models we could calculate a log N versus log S curve on the basis of assumptions (2) and (3). The third curve shown in Fig. 7.8 is that for the steady-state model. It begins with a slope of $-1·5$ and becomes progressively flatter. The big-bang models give similar curves. Thus it would mean that if we abandon Euclidean geometry in favour of the non-Euclidean geometries of the expanding cosmological models, the gap between theoretical prediction and observations widens still further. The gap can be plugged in the case of the big-bang models by asserting that n increased in the past. Indeed close agreement between theory and observation can be obtained in this way by requiring n to be larger in the past. This latitude is denied to the steady-state theory however. As we saw in Chapter 4, this theory requires the Universe to be unchanging over long periods of time. So n cannot increase either in the past or in the future. For this reason it was claimed by the Cambridge radio-astronomers that the radio-source count disproves the steady-state theory of the Universe. To what extent is this claim justified?

The earlier Cambridge surveys in the late 1950s had predicted even steeper slopes for the log N versus log S curves. However, these steep slopes turned out to be due to a number of experimental errors. The slope of $-1\cdot8$ arrived at in the early 1960s is believed to be more reliable, and thus presents a real challenge to the steady-state theory. A similar survey in the late 1950s made by Mills, Slee, and Hill in Australia gave a less steep slope, but still a slope steeper than $-1\cdot5$. Nevertheless, over the years considerable work has been done on this test, both on the observational and the theoretical sides, and in the light of information available today it does not appear that the test has any cosmological significance. The main points of these investigations are as follows.

The two surveys from Cambridge and Australia mentioned above were made at frequencies of 178 MHz and 408 MHz. In 1968 Bolton, Shimmins, and Wall in Australia made a survey at a much higher frequency—that of 1400 MHz. The slope of the log N versus log S curve in this survey was only $-1\cdot4$. The difference between this relatively flat slope and the steep slope of $-1\cdot8$ could be understood to some extent in terms of radio-source spectra. Fig. 7.9 shows the intensity versus frequency graph of three typical sources A, B, and C. A has a steep spectrum, with intensity falling in inverse proportion to frequency, while in B the spectrum is flat, that is, the intensity does not change much with frequency. Source A will therefore appear much fainter in a survey at 1400 MHz than at 408 MHz, whereas for B this effect will not be present. On the other hand, the source C, which has a rising spectrum over this range of frequencies, may appear faint at 408 MHz and bright at 1400 MHz. Thus it is essential to take into account the spectra of sources in interpreting the source count. The faintness of a source may not just be due to its distance away.

The assumption (3) (p. 216) of constant luminosity is a very dubious one. In general, sources vary considerably in their total output of radiation. Thus a source with a low value of S may either be a strong but distant source or a weak but nearby source. How can this be decided? If the object is identified with a galaxy and its red-shift measured then that, with the help of Hubble's law, gives information about its distance. So on this basis an increase in

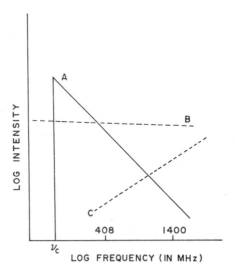

Fig. 7.9.   Simplified plots of radio-source spectra. An actual source
may have a combination of these. The steep spectrum A
terminates at a low enough frequency, where synchrotron
self-absorption sets in (see Box 7.6).

red-shift should be related to a decrease in flux level, if the latter is an
indicator of distance. A plot of red-shift against flux level is shown in
Fig. 7.10—and it does not reveal any correlation. Another confirma-
tion of this result is found in the comparison of red-shifts in the
previously mentioned Cambridge survey (called the 3C survey) and a
subsequent one (4C). The 4C survey picks out fainter objects than the
3C. However, the red-shifts of optically identified objects in the 4C
catalogue are not significantly higher than the red-shifts of such
objects from the 3C survey catalogue. What this seems to imply is
that, with his more sensitive instruments, the radio-astronomer does
not necessarily detect only very distant objects—he also detects
nearby *weak* sources of radiation.

We now come to another important point. When counting radio
sources, it is necessary to distinguish between the sources identified
with galaxies and the sources identified with QSOs. QSOs differ from

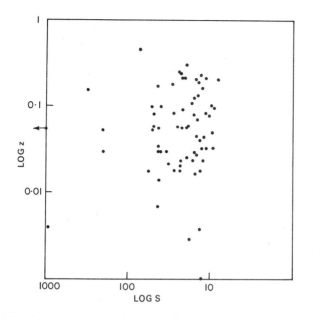

FIG. 7.10. (Drawn after Burbidge and Hoyle (1970), *Nature*, Lond. **227**, 359.)

radio galaxies in many respects, and it seems unreasonable to lump the two together. So sources in the 3C catalogue have been divided into three categories: (1) radio galaxies, (2) QSOs, and (3) sources not identified so far with any optical object. The log N versus log S curve for sources in each category has been separately plotted, and this leads to some remarkable results. The slope of the curve for radio galaxies is not much different from -1.5, indeed the curve is a little flatter than the Euclidean value. Similarly the slope for QSOs is also not very different from -1.5. We have already seen that galactic red-shifts are not very high and so a slope of -1.5 for them is not cosmologically significant. But the slope -1.5 for QSOs is significant *if it can be taken to imply that their high red-shifts arise from the expansion of the Universe.* At such high red-shifts QSOs must be very distant, and hence in the steady-state Universe the slope of their log N versus log S curve should be considerably flatter than it is observed to be. There is, as we shall see, considerable doubt about the interpretation of QSO red-shifts, and so the issue is unresolved. If the QSOs are local objects their slope of −1·5 is entirely consistent with a uniform distribution in space. But in this case no cosmological conclusion can be drawn from them.

The sources in the third category, the unidentified objects, exhibit a steep slope of −2·5. An argument given by Sir Fred Hoyle (see

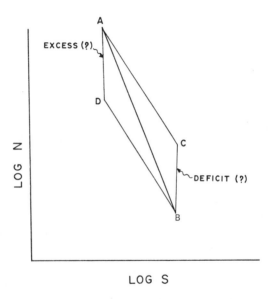

LOG S

Fig. 7.11.

Box 7.2 The unidentified sources

The log N versus log S curve of unidentified radio sources shows a slope of about −2·5, which is considerably higher in magnitude than the −1·5 slope expected in the Euclidean Universe. The steepness could arise in two possible ways, illustrated in Fig. 7.11.

Here AC and BD are two lines each with a slope −1·5 on the log N versus log S plot. The line AB has a slope −2·5. The points B and C represent the high-flux end of the 3C survey, corresponding to $S = 12·5$ units. The points A and D represent the low-flux end at $S = 5$ units. The observed curve AB implies that there are 10 sources observed at B and 93 at A.

Hoyle has argued that the survey indicates a deficit of sources at B.

If we follow the line AC we should have got $N = 23$ at C, instead of the observed $N = 10$ at B. The survey is over 3 sr (sr = steradian) in the sky. The deficit, according to Hoyle, is between 4 and 5 sources per steradian (the unit of a solid angle), and is not significant.

Ryle and his collaborators argue that we should compare AB with BD and not with AC. Thus at D we have $N = 40$, that is, an excess of 53 sources is observed over 3 sr. This, they claim, represents a significant excess over the predicted Euclidean curve.

Box 7.2) suggests that this is likely to arise from a paucity of nearby sources. A deficit of only about 4–5 sources per steradian (that is, about a twelfth of the total area of the sky) is involved. In contrast to this, the Cambridge group believes that the unidentified objects are very distant and that their steep slope is highly significant cosmologically, indicating that there were more such objects in the past than there are now. This issue cannot be resolved with the present data alone. It can only be settled if these objects are eventually optically identified and their red-shifts measured.

For these reasons the radio-source count has not turned out to be such a crucial cosmological test as it originally promised to be. The test is still potentially useful, but it must be applied with caution. Its application must wait until our understanding of the nature of radio sources, their optical identification, red-shift measurements, etc. have considerably improved.

Angular diameters

In 1959 Sir Fred Hoyle drew attention to a rather curious observable property of the non-Euclidean geometries used in cosmology. To understand this let us first consider what happens when we look at a galaxy or a radio source from successively increasing distances in a Universe which obeys the laws of Euclid's geometry. To simplify matters consider the spherical object shown in Fig. 7.12. When we look at it from a *distant* point O, it subtends an angle AOB $= \theta$ at O, where AB is the linear diameter of the object. Suppose AB $= d$ and the distance of O from the centre of the object is D. Then we have approximately,

$$\theta = \frac{d}{D},$$

provided θ is measured in radians.

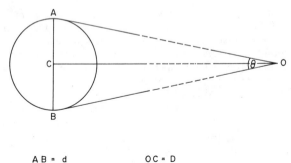

AB = d   OC = D

Fig. 7.12.

This means that if we increase D, θ decreases. This is shown by the straight line of Fig. 7.13(a), which is a plot of log θ against log D. The astronomer, looking at a distant object, measures θ directly, whereas his measurements of D and d are somewhat indirect, if they are at all

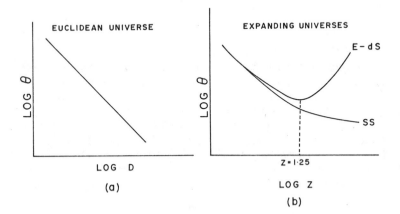

Fig. 7.13. (a) Represents the Euclidean case while (b) represents the non-Euclidean geometries of two cosmological models: steady-state and Einstein-de Sitter.

possible. Suppose we are looking at a lot of similar objects located at different distances. Then we should observe the relation shown in Fig. 7.13(a) between θ and D.

What happens in an expanding cosmological model? We have already seen that the red-shift z is an indicator of distance if we accept Hubble's law. So if we examine angular diameters θ of like objects of increasing red-shifts, what would we expect to find? Do we see θ decreasing to zero as we increase z to infinity? The answer, as calculated by Hoyle, was no. Two typical cases are described by the curves in Fig. 7.13(b). The curve labelled SS denotes the dependence of θ on z in the steady-state Universe. Here θ steadily decreases as z increases to infinity, but it does not tend to a zero value but to a fixed positive limit. In other words, an object located an 'infinite distance' away will still subtend a non-zero angle at the observer. The other case, the Einstein–de Sitter model (labelled E–dS), is even more startling. Here the angular diameter first decreases to a *minimum non-zero value* as z approaches the critical value $1·25$. Beyond this value θ begins to increase. Thus the more distant the object the bigger is the angle subtended by it at the observer, provided its red-shift exceeds $1·25$.

Qualitatively, we may understand this curious result in the following way. In the expanding Universe the angular separation between two galaxies as seen by us remains the same at all epochs; but their linear separation was less and less at earlier and earlier times. By contrast, the linear dimensions of a galaxy do not change with epoch. So the ratio of the linear diameter d of a galaxy and its distance from a nearby galaxy was larger in the past than it is now. This effect tends to increase the value of θ as we go further into the past, whereas the relation $\theta = d/D$ tends to decrease it. The overall result of the two effects determines the behaviour of θ. Different cosmological models predict different θ versus z curves.

This result could be a test of cosmological models provided we can apply it to a class of objects of fixed (or nearly fixed) linear size and large red-shifts. The latter requirement is met by QSOs, but the former is more difficult to satisfy. QSOs exhibit large variations in their intrinsic size. If later some subclass of QSOs turns out to be of nearly fixed linear dimensions this test can either be used to test different cosmological models or to test whether or not QSO red-shifts arise from the expansion of the Universe. The present status of this test is, however, uncertain.

Observations of the nearby Universe

The third category of test we shall call 'non-light-cone tests'; however, this is really a misnomer. Any observation of the Universe involves the use of electromagnetic radiation of some type and so, as in the second category, we are really looking at the Universe along the past light-cone. The difference between non-light-cone tests and the tests just discussed comes in the following way. The information sought here does not necessarily come from distant parts of the Universe, nor do the conclusions depend on the space–time geometry along the past light-cone. Thus our observations are of the nearby regions in the Universe, although the conclusions from them tell us about the state of the Universe in the remote past. This idea will be clearer with the discussion of the three tests described below.

The age of the Universe

If the Universe began with a big-bang, how long ago did this big-bang take place? This period, as measured by cosmic time, may be called the age of the Universe.

In the case of Friedmann models, we know the expansion factor as a function of the cosmic time. If we start measuring time from the big-bang, the present value of the time coordinate will be the age of the Universe. This calculation is a relatively straightforward one (see

---

†Box 7.3 The age and density of the Universe

In Friedmann cosmological models the expansion factor $S(t)$ satisfies the equation (see p. 123):

$$\left(\frac{dS}{dt}\right)^2 = -k + \frac{8\pi GA}{3S}. \tag{1}$$

Suppose we define t_0 as the present value of t, counted from the instant of the big-bang. The present value of Hubble's constant is given by

$$H = \left[\frac{1}{S}\frac{dS}{dt}\right]_{t=t_0}, \tag{2}$$

and the value of the parameter q by

$$-q = \frac{1}{H^2}\left[\frac{1}{S}\frac{d^2S}{dt^2}\right]_{t=t_0}. \tag{3}$$

Integrating eqn (1) it is possible to write the age t_0 of the Universe in terms of $T = H^{-1}$ and q. Fig. 7.14 shows the answer in a graphical form.

For the value $q = 1$ favoured earlier by Sandage this gives

$$t_0 = \left(\frac{\pi}{2} - 1 \right) T. \qquad (4)$$

The highest value of t_0 occurs when $q = 0$, giving $t_0 = H^{-1}$. The case $q = \frac{1}{2}$ represents the Einstein–de Sitter model. As mentioned in Chapter 3, this has $t_0 = \frac{2}{3} T$. The value of T is estimated at around 20 000 million years.

The parameter q can also be related to the mean density ρ of matter in the Universe, by the relation

$$\rho = \frac{3qH^2}{4\pi G}. \qquad (5)$$

For $q = 1$ this yields a value $\rho \simeq 10^{-26}$ kg m^{-3}, which is at least 10 times more than the density of matter seen in the galaxies, QSOs, etc. Where is the missing matter? Is it in intergalactic space? Is it hidden in black holes? Or is the value of q considerably smaller than unity? If, however, q is close to 0·1, there need not be any missing matter.

Box 7.3), and the result depends on the measured values of Hubble's constant and the deceleration parameter. Fig. 7.14 shows the variation of the 'age' of the Universe with q. Since we do not know q with any degree of confidence (see p. 214), it is not possible to fix the age

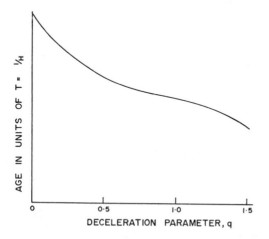

FIG. 7.14. The age of the Universe in units of $1/H$. The age decreases as q increases.

accurately. The maximum value is about 20 000 million years for $q = 0$. The value for $q = 1$ is about 11 500 million years.

The age of the Universe has played an interesting role in cosmological developments. With Hubble's original value of H, these above-mentioned figures were short by a factor 10. This was awkward to say the least, since even the age of the Earth, as estimated by the geophysicists, is about 4500 million years. This age discrepancy was one of the motivations behind the proposal of the steady-state Universe which, being without a beginning, did not have this age difficulty. By 1960, however, Hubble's constant had been revised, so that H^{-1} was about 10 000 million years. While this seemed to resolve the age discrepancy as far as the Earth was concerned, fresh problems had turned up. The age of the Galaxy, as estimated by Fowler and Hoyle, by two independent means using nuclear physics (see Box 7.4), gave a value of at least 13 000 million years, that is,

---

Box 7.4 The age of the Galaxy

The age of the Galaxy may be estimated in two ways. One method makes use of the ideas of stellar evolution discussed in Chapter 2. There we noticed that a typical star spends most part of its life in the main sequence. Thereafter, it branches off to the right on the Hertzsprung–Russell diagram. The rate at which the star covers these tracks on the diagram depends on its mass. Suppose now that we have a cluster of stars, all of which have branched to the right and are lying fairly close together on the Hertzsprung–Russell diagram. Then these stars must be approximately of the same age, which may be called the age of the cluster. The Galaxy must be at least as old as the oldest cluster found in it.

Using this method the ages of various clusters in the Galaxy have been estimated and the figure of 12 000 million years has been arrived at as some kind of an average. Clusters considerably older than this are believed to exist, but detailed calculations are needed to confirm this result. So the above figure should be taken as a conservative lower limit for the age of the Galaxy.

The second method makes use of the long time scales of radioactive decays of certain nuclei. For instance, the mean lives of thorium, ^{232}Th, and the uranium isotopes. ^{235}U and ^{238}U, are known, and they lie between 1000 million years and 20 000 million years. Suppose that the ratios of the abundance of these nuclei were known at the time of production and that these ratios are also known at the present times. Then, since the nuclei decay at different rates, we can compute the time taken from production to the present epoch. The production of, these nuclei is believed to take place in supernovae by addition of neutrons (see p. 49). So theoretically their production ratios may be estimated and compared with present observed ratios in the neighbourhood of the Sun. In this way the decay times were estimated by Fowler and Hoyle as about 15 000 million years.

The fact that both the estimates yield results of the same order is encouraging, since the two methods are quite unrelated.

greater than the age of all Friedmann models. At that time some 'pro-big-bang' theoreticians looked for possible loop-holes in the galactic-age estimates and sought to prove that the Galaxy was no more than about 6000–7000 million years old. This nicely fitted with a picture in which galaxies were formed soon after the big-bang and so had ages comparable to, but not exceeding, the age of the Universe. However, the current value of Hubble's constant has thrown this theory out of gear. For now the Universe seems to be sufficiently old to accommodate galaxies as old as those obtained by Fowler and Hoyle. Indeed it is now difficult to see why in a Universe as old as, say, 13 000 million years the galaxies should be no older than 6000–7000 million years. In other words, why was galaxy formation delayed for so long after the big-bang?

It is true that in the steady-state theory the Universe, being infinitely old, could accommodate galaxies of arbitrarily large age. All the same, it would be embarrassing for the theory if only very old galaxies are found in the Universe. Fig. 7.15 shows a graph of the expected age-distribution of galaxies in a large region (of the order of, say, 100 Mpc in linear size). The curve (an 'exponential' one) shows a decrease in the expected number of galaxies as they grow old. This happens because galaxies are all moving away from each other, so that the relative spacing of older galaxies is much larger than that of younger ones. The galaxies of 'zero' age should predominate in such a distribution. However, in practice, very young galaxies may not be observable, because the stars in them have barely formed. So there may be a 'threshold' in age, above which galaxies become observable. Also, galaxies may form in big groups rather than individually. If such groups or clusters remain bound for a while, in spite of expansion, galaxies of roughly the same age may be seen in a certain region of the Universe. These points, which can only be clearly defined after we have a theory of galactic origin and evolution (like that for stars), must be borne in mind when interpreting cosmological data. Broadly speaking then, in the steady-state model we expect to see galaxies of varying ages, younger galaxies being more common than older ones. By contrast, in the big-bang models there is probably a unique epoch when galaxies could form, and so galaxies should be predominantly of one age.

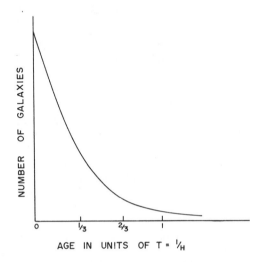

FIG. 7.15. The expected age-distribution of galaxies in a steady-state Universe.

The microwave background

We come now to what may be the strongest evidence of all against the steady-state model.

In 1965 A. A. Penzias and R. W. Wilson, two scientists from the Bell Telephone Laboratories in the United States, detected a background radiation in the microwave region (at a wavelength of about 7 cm). This discovery, like many other important discoveries in astronomy, came about accidentally. Penzias and Wilson were, in fact, making radio-astronomical measurements using an antenna originally designed to receive signals reflected from the 'Echo' satellites. The radiation they measured at 7 cm wavelength could be mostly accounted for by other sources such as atmospheric emission and ground emission. However, some residual radiation remained unaccounted for; and, more important, it was isotropic in character.

That is, it did not depend on any particular direction, implying that its source or sources, if any, did not exist close by (for example, in our Galaxy).

That such radiation should exist did not come as a surprise to the theoreticians, who already had a possible explanation for it. You will remember that, in the big-bang theory, a few moments after its origin, the Universe is dominated by radiation rather than matter. The radiation is that of a black body (see Boxes 2.3 and 4.8) and has a temperature in excess of 10 000 million K (10^{10} K) one second after the big-bang. However, as the Universe expands, the temperature drops sharply, and it was estimated by George Gamow in 1948 that it would be only a few degrees at the present epoch. At this temperature the maximum radiation occurs in the millimetre range of wavelengths. So the detection of radiation at a few centimetres wavelengths was not surprising. If the radiation observed by Penzias and Wilson is indeed a relic of the big-bang, then its temperature can be estimated. This turns out to be about 3 K.

How can we make sure that the radiation is indeed of a black-body type? This can be done by measuring the intensity at various wavelengths. The group at Princeton University, headed by R. H. Dicke, who in 1964 had independently arrived at Gamow's prediction, took the initiative in further measurements and obtained another point at a wavelength of 3 cm. The present status of the observations in relation to the black-body curve is shown in Fig. 7.16. The agreement on the longer-wavelength side of the peak is very good. The peak itself lies near 1 mm, corresponding to a temperature of 2·7 K. Measurements close to this wavelength and at shorter wavelengths cannot be obtained from ground-based astronomy. Instead we must use balloons and rockets to make measurements above the atmosphere. The data are therefore less accurate in this range.

The measurements indicated by letters CN correspond to indirect observations based on transitions in cyanogen (CN) molecules. When the CN-molecule is non-rotating, it has an absorption line at 3874·6 Å. In the first rotational state the wavelength is 3874·0 Å. When radiation of appropriate wavelength in the millimetre range is present the molecule makes transitions between the two states—the relative probabilities of the states being determined by the intensity of the radiation present (see p. 200). These probabilities can be estimated by measuring the relative strengths of the two lines. In this way the radiation intensity can be estimated in the neighbourhood of several stars in our Galaxy. This intensity is not inconsistent with the black-body curve.

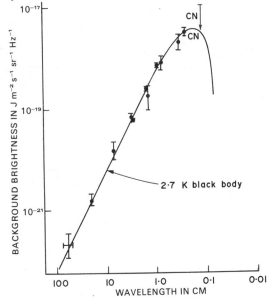

FIG. 7.16. Background radiation in the Universe. The observed points plotted together with a black-body radiation curve for a temperature of 2·7 K.

There was considerable confusion about direct measurements in this range. Earlier observations seemed to suggest radiation considerably in excess of the black-body type, but now most such claims have been withdrawn.

It is very important, from the cosmological point of view, to determine the exact nature of this background radiation at all relevant wavelengths. If it should turn out that the radiation is of a black-body nature, it would provide very good support for the big-bang models.

How has steady-state cosmology faced up to this challenge? It clearly cannot rely on the 'relic-radiation' interpretation, since the steady-state Universe never had a highly dense phase. The theory must provide other explanations, and here attempts have been made to ascribe the radiation to discrete sources all over the Universe. The argument goes as follows. First, it is noted that the total intensity of the microwave background is comparable to the energy density of (1) star-light in the Galaxy, (2) cosmic rays, and (3) the galactic magnetic field. However, whereas (1), (2), (3) are processes currently occurring in the Universe, the microwave background owes its origin to the big-bang, which happened long ago. Why should its energy density turn out to be comparable to these other processes? It is more reasonable to say that the background also arises from the processes presently occurring in the Universe. Next, we find that there is an isotropic background at other wavelengths, for example, in radio waves, visual light, X-rays, and γ-rays. But in all these cases we look for sources of radiation rather than argue in terms of the big-bang. Thus on both these counts the big-bang origin of microwave radiation looks something of an odd coincidence.

The explanation in terms of discrete sources requires a large number of sources emitting in the infrared and the microwave regions. Because of the red-shift, light from distant sources undergoes an increase in wavelength, thus transforming infrared radiation to microwave radiation. We have already seen (on p. 82) that Seyfert galaxies are strong infrared emitters. Perhaps with the growth of infrared and microwave astronomy (which is still in its infancy) we may discover a lot more infrared emitters, sufficient to explain the microwave background.

This explanation, however, has one serious difficulty. A source distribution in the sky will tend to introduce patchiness in the radiation—unless the number of sources is very large. There is very little patchiness in the observed microwave background, indicating that the number of sources must be at least as many (perhaps even 10 times more) as the number of ordinary galaxies, although individually the sources could be weaker. So far no such sources have been seen with any degree of profusion. Clearly future developments of infrared astronomy are important to settle this issue.

The remarkable isotropy of the background radiation poses problems even for the big-bang models. In the early stages of the big-bang Universe the regions of communication are terribly restricted by the so-called 'particle horizons' (see Box 7.5). Thus we

---

†Box 7.5 Particle horizons and the mixmaster universe

Suppose we ask the following question with regard to the Universe: 'What is the distance of the most remote galaxy which could be seen by us at the present epoch?' If the Universe were static and created a time t_0 ago, the answer would have been ct_0 where c is the velocity of light. Light from galaxies further than this has not yet had time to reach us. In a Friedmann Universe the calculation is a little more involved, but is based on the same principle. The answer is dependent, among other things, on t_0, the age of the Universe. This limiting distance represents the 'particle horizon' of the Universe. If the answer turns out to be infinite, as in the steady-state model, the Universe is said to have no particle horizon.

Suppose we now consider the particle horizon for the radiation Universe mentioned in Box 4.8. This Universe does have a particle horizon, in the very early stages, when the cosmic time t is measureable in seconds, the particle horizon has a radius of the order of light-seconds. That is, any given point in the Universe is in communication with a very limited region around it. To put it another way, a few seconds after the big-bang one part of the Universe could not know what another part of the Universe, located light-years away, is like. How then did the Universe manage to achieve such a high degree of homogeneity? In particular, how did the radiation background manage to have the same intensity over regions well out of communication with one another?

In the late 1960s Misner and his collaborators argued that the Universe may not have been isotropic to start with. Suppose we have a Universe in which there is no particle horizon in one direction, say the x-direction, but there are horizons in the other two directions. Then, in a limited set of directions around the x-axis, information can be transmitted over unlimited distances in the short time available. However, this will only partially solve the problem. To achieve homogeneity the transmission of information must take place in all directions. To this end, the 'Mixmaster Universe' was invented. In this the direction along which no particle horizon exists is not always directed along the x-axis—but is continually changing at random. In this way mixing of information could take place in all directions, leading to a homogeneity of radiation.

Although it has many interesting geometrical properties the mix-master Universe does not seem to serve the purpose for which it was invented. It does not seem able to mix and homogenize the radiation in all directions. So the alternative left to the big-bang cosmologists seems to be to assume *ad hoc* that the Universe was created in a homogeneous state.

should expect patchiness and anisotropy to be introduced at this early stage. Why then has the background remained so uniform? How, in other words, did parts of the Universe out of communication with one another manage to transmit information about the intensity of background in their own region so that a uniformity could somehow be achieved? Unless this could be shown explicitly we have to conclude that the Universe and the background radiation were created isotropic in the first place. While this is a possible assumption, to make it adds one more important but 'unexplained' item to the astronomer's shelf. A valiant but unsuccessful attempt to provide an explanation was made by C. W. Misner and his group at the University of Maryland, U.S.A. in the late 1960s. This work is briefly described in Box 7.5.

The origin of the elements

We have already looked at the current ideas on the origin of the elements in Chapters 2 and 4. Here we will be concerned with how our knowledge of the distribution of the elements can be used to test cosmological models.

Present calculations suggest that the bulk of the elements found in the Universe can be synthesized in the stars. The lighter nuclei like helium and deuterium can be made both in the stars and in the big-bang. The question is, 'Is the big-bang essential for these elements?' If it should turn out that the stars alone cannot manufacture enough of these nuclei then one is forced to take the big-bang origin of the Universe seriously.

Deuterium (the heavy-hydrogen nucleus) has been looked for in recent years in various parts of the Galaxy. In the direction of the Galactic centre, the ratio of D (deuterium) to H (hydrogen) is estimated in the range of 3–50 parts in 100 000. In the Orion Nebula the ratio is thought to lie between 1 and 100 parts in 10 million (this is deduced from the radio emission lines of DCN molecules). The present calculations, which depend on a number of theoretical assumptions and parameters, indicate that both supernovae and the big-bang Universe are able to manufacture enough deuterium to account for the observed ratios. The theory of supernova origin, however, is more restricted in the choice of parameters and may have to be abandoned if these parameters prove unrealistic.

A somewhat similar situation exists with regard to the ^{3}He nucleus which could also be synthesized in the big-bang as well as in supernovae.

Calculations for the big-bang origin can be related to the microwave background because this background, if of cosmological origin, provides information of radiation density in the very early stages of the Universe when element synthesis took place. There is at present some discrepancy, which may be summarized as follows. The abundances of the light elements (for example, the helium/hydrogen ratio) depend critically on the matter density and the rate of expansion of the Universe in its early stages. It turns out that the expansion rate in those critical seconds, based on Sandage's value of $q = 1$, is not fast enough to account for the observed abundances. Big-bang models with q close to zero are better placed in this respect. Some theoreticians have proposed alternatives to the Friedmann models. For instance, Dicke has suggested models obtained on the hypothesis that the gravitational constant has been steadily decreasing with epoch (we have discussed this possibility on p. 172). Hawking and Tayler have suggested an anisotropic big-bang, to speed up the overall expansion rate.

Calculations like these indicate the role of nuclear physics in cosmology. If the Universe did originate in a big-bang, violent activity in the form of nuclear reactions must have left its mark. The abundances of the various light nuclei provide a check on the possible conditions existing in such a big-bang. Needless to say, any spectacular advances in our knowledge of nuclear and elementary-particle physics may drastically alter the present picture.

The nature of QSO red-shifts: an unsolved problem

It must be clear from the preceeding discussion that QSOs will play a decisive role in cosmology, provided their red-shifts can be established as due to the expansion of the Universe. The QSO red-shifts at present represent a challenge to the theoretician. In this section we will consider the main points in the controversy about the nature of these red-shifts.

Red-shifts in QSO radiation can be thought of as being due to any one of the following effects: (1) the expansion of the Universe, (2) the gravitational field in a QSO, and (3) the Doppler effect. Here I want to weigh the pros and cons for each explanation. It must be remembered, however, that these effects are not mutually exclusive—the red-shift in an object can arise from some or *all* of these causes. To be of 'cosmological significance', however, it must be clearly established that the causes (2) and (3) contribute only a negligible fraction towards the total red-shift. Otherwise, we cannot use the QSOs as distance indicators in cosmology. For example, if a QSO has a red-shift $z = 2$, we can assert only that it is further away than a galaxy of red-shift $z = 0.2$, *provided* we are sure that most of the QSO red-shift is due to the expansion of the Universe.

The present evidence is largely of a negative character. That is, it raises objections to each one of the above-mentioned effects, and it is a case of choosing the least objectionable explanation. If the objections are insurmountable, then we must search for a new effect. These choices to some extent involve subjective judgements and prejudices, and this is largely why the present controversy exists.

Let us consider the cosmological explanation first. The first objection to this has already been mentioned on p. 215: the Hubble diagram of QSOs shows no clear-cut red-shift–magnitude relation. The very idea of the expansion of the Universe came from Hubble's linear law for galaxies, and since no such law has been found for QSOs, the case for their red-shifts arising from the expansion of the Universe is weakened considerably. The defendants of the cosmological explanation argue that the enormous scattering in the Hubble diagram is due to large variations in the luminosity of QSOs. This could be true, but then QSOs are useless for applying the red-shift–magnitude test. Recently attempts have been made to minimize the scatter by choosing special classes of QSOs. For instance, we can consider QSOs that are radio emitters with flat spectra (see Fig. 7.9 for radio sources), QSOs that are radio emitters with steep spectra, and so on. The hope is to achieve for QSOs what Sandage did for galaxies, that is, to select a class of similar objects for which there is little luminosity variation, so that their Hubble diagram will indicate a linear relation. While such attempts have led to some 'improvement' of the Hubble diagram, the effect in QSOs is in no way comparable to that of galaxies.

Another objection to the idea that the red-shift arises from the expansion of the Universe is a more theoretical one. If a QSO subtends a certain angle θ (see Fig. 7.12) at an observer O from a distance D, the linear size d of the QSO is given by $d = D\theta$, where θ is in radians (not degrees). Suppose the QSO also shows a rapid variation in its optical or radio output, or in both. Let T be the time scale associated with such a variation. Then the linear size d is limited by the fact that it cannot exceed the distance travelled by light in time T (this requirement is imposed by the special theory of relativity). Hence if c is the velocity of light, $d < cT$. Combining this with the earlier relation we get

$$D < \frac{Tc}{\theta}.$$

If T and θ are known, we can check whether the limit on D is consistent with the cosmological explanation of the red-shift. Of these, T is observationally determined, while θ may either be measured directly from observations or may be estimated from a theoretical argument. I shall describe the latter method for obtaining θ.

In Chapter 3, we looked at the synchrotron emission process for radiation, which is believed to take place in most strong radio sources. Here an electron accelerated by a magnetic field radiates. Now it so happens that at a certain critical frequency the source begins to re-absorb the radiation emitted by it. This is known as 'synchrotron self-absorption'. This critical frequency depends on the size of the source, its magnetic field, and the flux radiated at that frequency (see Box 7.6). The spectrum of the source shows a sharp

---

†Box 7.6 Self-absorption of radiation

Suppose we have an enclosure in which there are sources of radiation and suppose that ideally the radiation does not leak out of the enclosure. Then such a system tries to attain an equilibrium condition when a certain amount of re-absorption of radiation by the sources takes place. The 'black body' represents a state where this equilibrium has been attained.

In a radio source with a steep spectrum (see Fig. 7.9) arising from synchrotron radiation, the intensity of radiation seems to keep on rising at lower and lower frequencies. This process cannot go on for ever—because this would lead to unlimited intensity of radiation in the source. At some stage the absorption process must set in, which re-energizes the charged particles at the expense of radiation. The situation at these frequencies then begins to look like that in a black body.

The frequency at which this begins to happen is called the 'critical frequency'. This is given by the following approximate formula

$$\nu_c \simeq 2.85 \times 10^{10} \, H^{\frac{2}{5}} \, S^{\frac{2}{5}} \, \theta^{\frac{4}{5}} \text{ MHz}$$

where H is the magnetic field strength in units of gauss (1 gauss = 10^{-4} tesla) θ is the angular diameter of the source in seconds of arc, and S is the flux density from the source.

Thus for $\nu < \nu_c$ the spectrum shows a sharp drop due to the effects of self-absorption. In practice we know ν_c and S and can estimate H from other factors (see Box 3.5). Hence we get an estimate of θ.

drop in the radiation at frequencies lower than the critical frequency, and so tells us where self-absorption begins. In this way it is possible to estimate the size of the source. When these considerations were applied to sources like 3C-273, for which time variation in luminosity is found, the results were, in general, not favourable to the cosmological interpretation of the red-shift. That is, the angular size required by the synchrotron-radiation argument is too large, and consequently the limit on D too small, to admit the possibility that these sources are at cosmological distances.

There is a loop-hole in this argument, which was pointed out by Martin Rees in 1966. If a QSO is expanding with a high velocity, the limit on d set by time variation is considerably increased. It is in fact increased by the factor

$$\gamma = 1 \bigg/ \left(1 - \frac{v^2}{c^2}\right)^{\frac{1}{2}},$$

where v is the velocity of expansion and c is the velocity of light. For v close to c, the γ-factor can be very high. If, however, such rapid expansion is not possible for other reasons, the only alternatives to abandoning the cosmological interpretation of the red-shift are either to abandon the synchrotron-radiation idea or to think of more complicated models. This is being done.

In 1966 another test was made of the cosmological hypothesis, using the source 3C-9 with a red-shift of 2. The following assumptions were made: (1) the source is a distant one so that its red-shift is due to expansion of the Universe; (2) the intergalactic medium is made of neutral hydrogen—the simplest material imaginable and considered most plausible by cosmologists. Now neutral hydrogen emits and absorbs light of wavelength 1216 Å, more commonly known as Lyman-α. Let us consider light of a wavelength λ, shorter than 1216 Å, coming towards us from the source. In the light's passage through the expanding Universe the wavelength gradually increases until it reaches 3λ at the Earth. If 1216 Å lies in between λ and 3λ, this light will attain the wavelength 1216 Å somewhere on the way and will therefore be absorbed—the amount of absorption depending on the density of the intergalactic hydrogen. This means that the continuum spectrum on the blue-ward side of Lyman-α (that is, on the short-wavelength side) should show an appreciable drop in intensity. In fact no such drop is seen. This can be interpreted *either* by saying that the QSO is not distant and therefore that no absorption

240   THE STRUCTURE OF THE UNIVERSE

is expected en route *or* by arguing that the intergalactic medium is not made up of neutral hydrogen. This latter hypothesis finds favour with those who do not want to give up the cosmological interpretation of the red-shift.

There have been attempts to establish the cosmological nature of QSO red-shifts by identifying the presence of QSOs in the neighbourhood of galaxies with similar red-shifts. The argument is that if we find a QSO situated very close to a galaxy and both have a red-shift of, say, 0·2, then we *know* that the QSO red-shift must be cosmological because that of the galaxy certainly is. The proximity is established by statistical arguments. That is, given the angular separation between the QSO and the galaxy, the probability is calculated that the two should be found so near on a random (that is, pure chance) basis. If the probability is very low, say not exceeding 1 per cent, then the association between the QSO and the galaxy is very likely to be genuine. Several such associations have been demonstrated recently, apparently lending support to the cosmological idea.

Geoffrey Burbidge and Steve O'Dell have criticized this approach for the following reason. By selecting a QSO–galaxy pair with small red-shifts the observer is making a prejudgement in favour of the cosmological interpretation of red-shifts, because he knows that galaxies do *not* have high red-shifts, whereas the QSOs do. Suppose we select QSO–galaxy pairs without reference to red-shifts, and then see whether the two are associated in the statistical sense, and *then* measure the red-shifts. If the cosmological idea is right the red-shifts should turn out to be approximately the same.

Recent work along these lines is somewhat disturbing for the supporters of the cosmological red-shift interpretation. *The red-shifts of QSOs and galaxies associated in this way have generally turned out to be very different.* The most striking case to date is not of a QSO–galaxy pair but of a pair of QSOs. In 1973, Hazard, Wampler, Baldwin, Burke, and Robinson found that the radio source 4C-1150 identified with the optical source 1548+115 (see Fig. 7.17) consists of two QSOs of 17th and 19th magnitude with red-shifts 0·44 and 1·90. If these red-shifts are of cosmological origin, one QSO must be 4–5 times further away than the other. Yet on the sky they are only 5″ of arc apart. On a chance basis the probability of these two QSOs lying so closely in the same direction is 6 parts in 1000. That they are physically associated is also indicated by many common spectral

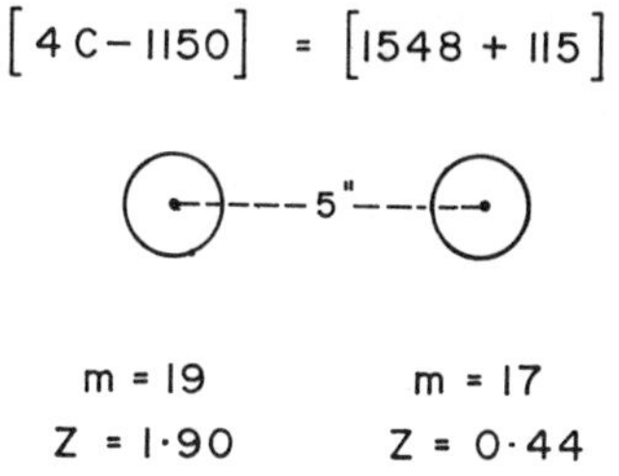

$$\left[\,4\,C-1150\,\right]\; =\; \left[\,1548 + 115\,\right]$$

$$\begin{array}{cc} m = 19 & m = 17 \\ Z = 1{\cdot}90 & Z = 0{\cdot}44 \end{array}$$

Fig. 7.17.

features. If more cases like this turn up the cosmological interpretation of red-shifts will have to be discarded, because these cases attack the very heart of the hypothesis that sources at a given distance should have the same red-shift.

The Doppler-shift explanation of red-shifts in QSOs was first advanced by J. Terrell in 1964. He suggested that QSOs were comparatively small objects ejected from the nucleus of our Galaxy at large velocities. This explanation was not taken up by other astronomers because (1) there does not seem to be any explosion occurring in the centre of our Galaxy which could throw out such objects and (2) the total energy required for ejecting so many QSOs would be simply beyond the capabilities of our Galaxy—it would break up with the violence of the event. Terrell's idea was revived in the following year by Hoyle and Burbidge in a more acceptable form. They argued that QSOs could have been ejected from nearby exploding objects. In particular they suggested NGC 5128 or Centaurus A (see Fig. 3.18, p. 78) as a possible source. As a rule, these sources and the QSOs ejected by them may be located within distances of the order of 10 Mpc (as opposed to 1000 Mpc or more in the cosmological interpretation of QSO red-shifts). This gets round the two objections against Terrell's idea. The exploding objects do seem to be seats of enormous output of energy and may be capable of ejecting several QSOs. It must be remembered here that the nearer we assume a QSO to be, the less energy it has to be endowed with. Because, in order to produce the observed brightness at the Earth, the required intrinsic brightness of the QSO increases roughly in proportion to the square of its assumed distance from the Earth. Thus a QSO under this local interpretation of the red-shift need not be as powerful as it would be under the cosmological interpretation.

A strong objection to this idea, however, comes from another quarter: the apparent lack of blue-shifts. Fig. 7.18 shows two

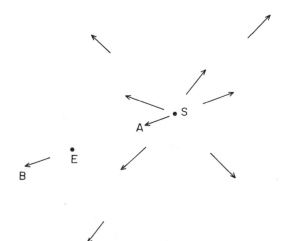

FIG. 7.18. E represents the Earth; S represents the source of the explosion. A and B are QSOs sent out in the direction of the Earth.

examples under the Burbidge–Hoyle scheme. E is the Earth and S the source of explosion. A and B are two QSOs sent out in the direction of the Earth. Of these A, which lies between S and E, must be blue-shifted while B, which has gone beyond E, is red-shifted. Why do we not see any blue-shifted QSO? This may be countered by saying that if the explosion occurred long ago all objects directed towards the Earth (or our Galaxy) may have crossed it and gone beyond it, like B in the above example. While this could be true for nearby seats of explosion it is less plausible for more distant sources. If we take the case of an isotropic explosion (that is, the case where objects are ejected by S equally in all directions) and assume that no objects (like B) have yet gone past the Earth, then blue-shifted objects are expected to predominate over red-shifted ones in a typical survey of QSOs up to a specified level of observed brightness.

The reason for the lack of blue-shifts may be that there are strong selection effects which conspire against the detection of blue-shifts. For instance, the continuum radiation of 3C-273 shows an increase in the infrared region. Now 3C-273 has a small red-shift ($z = 0.158$). If, however, we have another QSO with a large blue-shift and similar excess radiation in the infrared, that part of the continuum spectrum will be shifted to the visible range, leading to a bright continuum. On the other hand there are no known strong lines in the infrared which, on a similar blue-shift, could stand out against this continuum. Thus large blue-shifts may be inherently difficult to measure. Still, the fact that *no* blue-shift, however small, has yet been reported argues against this theory.

Finally, we come to our second suggestion about the origin of QSO red-shifts, that the red-shift may be of gravitational origin. Since QSOs give the appearance of massive concentrated objects it is natural to expect strong gravitational fields round them, and hence large gravitational red-shifts. Indeed, when the first two QSOs, 3C-273 and 3C-48, were discovered, Maarten Schmidt did examine the possibility of their red-shifts being of gravitational origin. With an elegant but technical argument he was able to demonstrate that this hypothesis would be untenable except under very stringent conditions. The main difficulty with this idea, in Schmidt's argument, comes from the fact that the spectral lines of QSOs are too narrow. In the gravitational red-shift explanation the red-shift from the surface, when it is small compared to unity, is given by

$$z = \frac{GM}{c^2 R},$$

where M is the mass of the object and R its radius (see p. 159 for the details of this calculation). Near the surface the density of matter is usually very low, so that the outer shells of the object contain a very small fraction of the total mass M. Now a spectral line will appear broad if it comes from a fairly thick shell near the surface of the object, so that there is a variation in the value of R. The line from the inner part of the shell will be red-shifted more than the line from the outer part, because in the above formula the value of R is less for the former than for the latter, whereas M is the same. So the combined effect is to give a broad line in the red-shifted spectrum. On the other hand, a narrow line implies a narrow emitting shell, and it becomes difficult to accommodate all the radiating matter in a thin shell.

In 1966 William Fowler and Sir Fred Hoyle got round this difficulty by proposing that the emitting region is in the centre of the object rather than on the surface. Light comes out through gaps in the absorbing chunks of cloud in the outer parts in this model. Qualitatively this can also explain why the emission-line red-shifts are different from absorption-line red-shifts in many QSOs. The reason is that emission takes place at the centre and absorption at the exterior of the object, and the gravitational fields are different in the two regions. The full implications of the Fowler–Hoyle model have yet to be worked out quantitatively to see whether it successfully meets all objections.

A qualitative argument against the gravitational-red-shift models is that QSOs with high red-shifts do not seem to show any high degree of compactness (or high density) compared to low-red-shift QSOs. Normally we would expect high-red-shift QSOs to be highly compact systems (close to black holes).

This has been a single and brief review of the problem of red-shift origins. Detailed models of QSOs present many more complicated problems, but here I have concentrated on the individual failings of each theory, in order to demonstrate that none of them is perfect. A particular failing of one theory may not present a problem to the others. For instance, the lack of blue-shifts, which poses such a major problem to the Doppler-effect theory, is explained naturally by the cosmological interpretation in which the expansion of the Universe gives rise to red-shifts only. Similarly, the red-shift–magnitude diagram will be a scatter diagram rather than a straight line according to the gravitational-red-shift–interpretation because under this hypothesis QSO red-shifts do not depend upon distance. However, a perfect all-embracing theory of this phenomenon is still to be found.

Concluding remarks

Cosmology has certainly developed maturity as a science since its infancy in the days of Hubble, Einstein, de Sitter, and Eddington. The wide range of models of the Universe which have been proposed over the years reflects the boldness of theoreticians in tackling the 'biggest' problem ever and their implicit faith in physical theories. And the many observational checks of the type discussed in this

chapter indicate how cosmology is now fulfilling the main requirement of a scientific discipline: that it should make testable predictions.

Nevertheless, it would be premature to assert that the cosmological problem is solved—that we know what the Universe is (or
was) like. The many controversies which still exist in this field bear
testimony to this fact. In this chapter I have avoided the temptation
of coming to definitive conclusions like 'the Universe had a big-bang
origin' or 'the Universe is in a steady-state'. In my view observational
astronomy has still a long way to go before we can draw such
conclusions. Nor are such conclusions independent of other issues,
like the nature of QSO red-shifts. In the final chapter we shall look
briefly at some of the outstanding issues now facing the physicist and
the astronomer. The resolution of any one of these problems should
advance our knowledge of the Universe by another big step.

8

Epilogue

The essential thing in science is for the scientist to think up a theory. There is no way of mechanizing this process; there is no way of breaking it down into science factory. It always requires human imagination, and indeed in science we pay the highest respect to creativity, to originality ... we do not honour scientists for being right; it is never given to anybody to be always right. We honour scientists for being original, for being stimulating, for having started a whole line of work.†

Sir Hermann Bondi

Astronomy, like other branches of science, has faced difficult problems from time to time, and out of their solutions have emerged important discoveries. The motion of planets posed a problem for several centuries until it was understood in terms of Newton's law of gravitation. Investigations of faint nebulae led to the concept of an expanding Universe. The mystery of the source of solar energy was finally resolved by our increasing knowledge of nuclear physics. And now the Universe has presented us with many more unsolved problems, some of which have been mentioned in this book. Here I shall summarize just a few of these problems.

QSO red-shifts

There are now nearly 500 QSOs known, with detailed observations of many of them. Yet, so far no satisfactory model of a QSO has emerged, or even a reasonable explanation of its red-shift (or red-shifts!) Apart from the three known causes of red-shifts discussed in the last chapter, does there exist another, a new one?

Solar neutrinos

In spite of great advances in the theory of stellar evolution, there is a problem right at our doorstep. The nearest star, the Sun, which we should know best, is not sending out neutrinos in the quantity

† From *Cosmology Today* (ed. L. John). B.B.C. Publications (1973).

expected theoretically. In Chapter 2 we saw that neutrinos are by-products of nuclear reactions taking place in the Sun. Yet attempts to detect these neutrinos on the Earth have so far proved unsuccessful. The discrepancy between theory and observation is serious enough to keep busy not only astronomers but also nuclear physicists and geophysicists. Are we still missing some important clue in nuclear physics or in the structure of the Sun?

Radio sources

What is the primary source of energy of the strong radio sources? This question is similar to the problem of source of solar energy which troubled the astronomers until the 1930s. So far no satisfactory solution has been advanced. An intriguing possibility not yet fully explored exists in the form of 'white holes', the opposite of black holes. These are pockets of space where matter with high energy is pouring *out* (rather than *in*, as in black holes).

Black holes

Do black holes exist? Have they already been detected? Do they account for a lot of 'unseen' matter in the Universe? The cosmological models described in Chapter 4 generally predict densities of matter in the Universe far in excess (that is, by factors 10–100) of that actually observed. It has been suggested that black holes may account for this discrepancy. These intriguing objects are of interest not only to the astronomers but also to mathematicians, who like to study the weird geometrical properties of space–time in and around black holes.

Gravitational radiation

How is this radiation interpreted theoretically? Has it been measured observationally? There are doubts on both counts, as discussed on p. 163. Potential sources of such radiation, if it exists, will be found in the violent phenomena of astronomy rather than in the small world of the laboratory physicist. The unambiguous detection of gravitational radiation will represent a major event—for it will break the monopoly of the electromagnetic radiation as the quickest information carrier in the Universe.

And so the list goes on. Perhaps astronomers of the next generation will have solved these problems while discovering new ones.

Are new laws of physics necessary?

So ends our brief survey of the Universe. Limitations of space and time have necessarily restricted the discussion to only some of the highlights. We will finish by considering the question raised above: Is astronomy telling us that what we know about basic physics is incomplete and needs augmentation?

To put the discussion in proper perspective let us first recall that it was astronomy that played a major role in the early development of physics. The first of the four fundamental interactions of physics—gravitation—was properly tested not in terrestrial laboratories but in the heavens. Over the last couple of decades astronomy is rediscovering that role. By presenting the theoreticians with phenomena on a vastly grander scale than could ever be achieved in a terrestrial laboratory, astronomy is initiating the application of physical laws to unknown territories. It is not surprising, therefore, if stumbling blocks are being encountered and will continue to be encountered. The question is, 'Are these difficulties genuine reflections of the inadequacy of physical laws or do they arise from uncertain observational data and limitations in the imagination of the theoreticians?' Answers differ, and I wish to record the various points of view.

First, it cannot be denied that there have been cases where a physcial theory has fallen victim to uncertain—even wrong—observations, which have been subsequently withdrawn. The most notable example of this is the steady-state theory of the Universe. This theory makes certain very definite predictions, and must be abandoned if these are not borne out by observations. The theory has been attacked, and even claimed to be disproved, from time to time, with the help of data which have either been subsequently withdrawn or proved inconclusive. I have described instances of this in the last chapter. So far the only genuine difficulty faced by the theory is that arising from the microwave background. If no alternative sources (to big-bang) of this background can be found, the steady-state theory must be abandoned. Yet the popular belief, shared even by most astronomers, is that the theory has already been disproved once for all.

The outstanding problems enumerated at the beginning of this chapter might on the face of it, tempt us to rush out and look for new laws to explain them. In the case of QSO red-shifts it may be possible to introduce another source of red-shifts by admitting that the masses of elementary particles are not constant but depend on space–time properties and the other matter in the Universe, in the Machian way (see p. 171). If the mass of an electron in a QSO is different from that on the Earth, the spectral lines in the QSO will be different in wavelengths from those on the Earth. It is similarly possible to change other 'constants' of nature and investigate the consequences.

Yet scientists are basically a conservative lot, in spite of some of their more radical pronouncements. They prefer to stick to what has been 'tried and proven' rather than open the floodgates to a tide of new and very often crazy ideas. Distinguished scientists of the last century, Lorentz, Michelson, Poincaré, Mach, and others, were hostile to the special theory of relativity when it was first proposed in 1905. To some extent this conservatism is quite justified. There would be no end to the new ideas once a free for all had been declared —and to investigate each new idea in great detail is simply beyond our present-day capabilities.

It may well be the case that some observations cannot be explained because the scientists have not investigated *all* the consequences of the existing laws. This happens in other branches of science. For example, the apparrently inexplicable behaviour of superconductivity was finally accounted for within the framework of the present-day quantum physics. However, this outlook can be carried too far and may lead to complacency. The scientists of a hundred years ago believed that all the basic laws of physics were known and only the details needed to be worked out. This outlook came in for two rude shocks at the beginning of the present century—the theory of relativity and the quantum theory.

More radical physicists today would tend to think that the present laws of physics are again in for some rude shocks, but this time from the astronomical world. As J. B. S. Haldane remarked, the Universe is 'not only queerer than we suppose, but queerer than we can suppose'. It may therefore be difficult to say precisely in what new directions astronomy will guide fundamental physics, but that the new leap forward will happen in the near future is very much on the cards.

I personally feel that the new major influence of astronomy will come about through the removal of the straight jacket placed on physics by the limitations of a terrestrial laboratory. Since laboratory experiments have guided the growth of physics over the last two centuries, physicists are accustomed to thinking in terms of 'local' laws of physics, that is, in terms of laws applicable 'here' and 'now'. The application of these laws to astronomy has been through a process of cautious extrapolation. *This hardly does justice to the grand laboratory provided by the Universe as a whole.* A more suitable outlook has been provided by Mach's principle (see p. 165) and the Wheeler–Feynman theory (see p. 190), which bring the Universe intimately into contact with 'local' phenomena. Once the physicist appreciates that isolation of the local environment from the Universe as a whole is not possible, he will be willing to make a change from the local outlook to the global one. Indeed, the unity of the Universe demands that while we study what is happening 'here' and 'now' we cannot, for consistency and relevance, ignore what goes on 'out there'.

Glossary

absolute magnitude See text (p. 16).

absolute space When Newton formulated his laws of motion he was faced with one conceptual problem. For measuring the velocity of a body it is necessary to know the background frame of reference. Newton postulated the absolute space as an abstract entity which provided this background. Thus all velocities and accelerations in Newton's laws are measured against the background of absolute space.

absorption spectrum The series of dark lines or bands which appear in the spectrum of a luminous object form its absorption spectrum. The darkening is due to absorption of light from the object at the characteristic wavelengths, produced by the surrounding medium. The absorption spectrum therefore tells us something about the nature of the absorbing medium. (See Box 2.4.)

aeon Unit of time equal to one thousand million years.

alternating current Electric current which periodically switches its direction. This arises from the to-and-fro motion of electrical charges.

Angström A length scale used for measuring the wavelength of electromagnetic radiation in the infrared to the ultraviolet range. One angström (Å) equals a hundred millionth part of a centimetre.

apparent magnitude See text (p. 16).

astrophysics Subject dealing with the physical properties of astronomical objects.

atomic clock Clock making use of the oscillations in atomic nuclei (such as the caesium atom). These have extremely high degree of precision, of the order of one part in 10^{11} or more.

3C catalogue Third Cambridge catalogue of radio sources. The Mullard Radio Astronomy Observatory of the University of Cambridge has conducted surveys of radio sources in which the sources are catalogued in a certain order of their directions in the sky. The Nth source is labelled 3C-N. Although fourth and fifth catalogues include sources much fainter than those in the 3C catalogue, the latter has the merit of being complete down to a specified level of faintness.

causality The principle stating that causes precede effects.

centrifugal force The radially outward force experienced by a body moving in a circular motion about a centre.

classical physics Physics describing natural phenomena without recourse to quantum theory.

colour index See text (p. 20).

colour filters Filters admitting light of specified range of wavelengths. These are used by the astronomer in his telescope, when he wants to study the properties of light emission from the object within a selected wavelength band.

continuum emission Emission of light over a continuous range of wavelengths. This contrasts with spectral lines which represent sharp peaks or troughs at discrete set of wavelengths. (See Box 2.4.)

cosmological principle The principle that the Universe looks the same in all directions and that it has the same properties at all points when viewed by any of a set of fundamental observers throughout the Universe. Thus the Universe, according to this principle, has no preferred position or a preferred direction.

cosmology Subject dealing with the large-scale structure, evolution, and the origin of the Universe.

Coulomb barrier According to Coulomb's law, like electric charges repel with a force inversely proportional to the square of their distance apart. This means that two such charges are prevented from getting very close to each other. This repulsive effect is expressed by saying that there exists a Coulomb barrier between the like charges.

deuterium Heavy hydrogen. A hydrogen nucleus contains only one proton whereas a deuterium nucleus contains one proton and one neutron.

eccentricity A mathematical parameter, often denoted by the letter e, specifying the nature of a planetary orbit. Thus $e = 0$ corresponds to a circular orbit. As e increases from 0 to 1 the orbit becomes more and more elliptical. $e = 1$ represents a parabolic orbit while higher values of e describe hyperbolic orbits. Planets are known to have only elliptic orbits.

electrodynamics Subject dealing with motion of electrical charges under the influence of electric and magnetic fields.

electron Negatively charged particle forming a constituent of an atom. The electron moves in orbits round a central atomic nucleus. Free electrons also exist. The mass of an electron is $9 \cdot 108 \times 10^{-31}$ kg and its electric charge is $4 \cdot 802 \times 10^{-10}$ electrostatic units.

electronvolt (eV) Work needed to be done to move an electron against an electrical potential barrier of one volt. In terms of the conventional energy unit, the joule, 1 eV equals approximately $1 \cdot 602 \times 10^{-19}$ joules.

elementary particles Particles believed to be the primary constituents of matter. Experiments over the last two decades have revealed the existence of more than a hundred such particles.

emission spectrum The series of bright lines in the spectrum of a luminous object. The lines represent packets of energy emitted by excited (i.e. energetic) atoms or molecules in the object at characteristic wavelengths. (See Box 2.4.)

entropy A measure of disorder in a physical system. The law of increase of entropy means a physical system goes from order to disorder.

epoch A typical moment of time in the history of the Universe.

Euclidean geometry Geometry based on Euclid's postulates.

event horizon The boundary of a space–time region such that events taking place beyond it will never be visible to a given observer.

frame of reference This is needed to specify the location and timing of a physical event.

frequency The number of times a repetitive event occurs in a given unit of time.

Friedmann Universe models Mathematical models of the Universe first worked out by the physicist A. Friedmann in 1922 using Einstein's general theory of relativity. These models require the Universe to originate in a big-bang.

γ-*ray* Electromagnetic wave of very high frequency, frequency upwards of $\sim 10^{20}$ Hz.

ground state (of an electron) The ground state of a physical system is the state in which it has the lowest possible energy (See p. 199).

hertz Unit of frequency named after the nineteenth-century scientist Hertz who first produced and detected electromagnetic waves in the laboratory.

homogeneous Having the same physical properties at all points of space.

Hubble's constant The ratio of the apparent velocity of recession of a distant galaxy to its distance. Its present estimate is about 53 km s^{-1} per Mpc.

equilibrium The unchanging state of a system resulting from the balance between various competing forces.

inertia The property of matter characterizing its resistance to change of state (of rest or of uniform motion in a straight line).

inertial frame of reference Frame of reference in which Newton's first law of motion holds good, i.e. a frame with respect to which a body under no forces remains at rest or has uniform motion in a straight line (see p. 166).

ionization Splitting of neutral atoms or molecules into components carrying equal and opposite electric charges.

isotopes Atomic nuclei with the same number of protons but different numbers of neutrons. Thus deuterium is an isotope of hydrogen.

isotropic Having the same property in all directions.

λ-*term* A term introduced by Einstein to represent repulsion between any two pieces of matter separated by a long distance (on cosmic scale). This force was supposed to increase in proportion to their intervening distance, being negligible at distances of the terrestrial or interplanetary order. At present there is no strong reason, theoretical or observational, for retaining this term in Einstein's equations describing gravitation (See Box 4.9.)

light-cone Region in space–time diagram accessible to light emitted from a point. Signals travelling faster than light lie outside this cone while signals travelling slower than light lie within it.

light-year Distance travelled by light in one year. This is approximately 9×10^{12} km, i.e. nine million million kilometres.

light-second Distance travelled by light in one second, i.e. approximately 300 000 km.

luminosity The amount of energy emitted by an object in a given unit of time is called its luminosity. For example, the Sun has the luminosity of 4×10^{26} joules per second.

Maxwell's equations In the third quarter of the nineteenth century the Cambridge physicist James Clerk Maxwell derived a comprehensive set of equations to describe the electromagnetic theory. These are known as Maxwell's equations.

Minkowski Universe Static (i.e. non-expanding) Universe in which the space has Euclid's geometry and in which the rules of Einstein's special theory of relativity apply.

neutrino An elementary particle of zero electric charge travelling with the speed of light.

neutron Electrically neutral elementary particle usually found in the nucleus of an atom. Its mass is slightly more than the mass of the proton and about 1838·65 times the mass of the electron.

neutron star Star composed of matter in the form of neutrons. Such stars have very high ·densities, a typical neutron star density being at least $\sim 10^{10}$ kg m^{-3}, some 10 million times the density of water.

NGC Catologue New general catalogue of astronomical objects visible in the visual range. Objects are referred to by their number preceded by the letters NGC. The original NGC catalogue was published by J. L. E. Dreyer in 1888. It has since been revised and augmented.

non-Euclidean geometry A geometry based on postulates differing from Euclid's in one or more respects.

nucleus Central dense region. The nucleus of an atom contains protons and neutrons. The nucleus of a galaxy may contain densely packed stars, gas and dust.

parsec A measure of distance slightly more than three light years. See text (p. 15). A megaparsec (Mpc) equals a million parsecs.

perfect cosmological principle (PCP) The principle according to which the Universe in the large is unchanging in time. The Universe is also expected to have the symmetries imposed by the ordinary cosmological principle.

photon Particle associated with light. According to quantum theory light consists of packets of energy; a packet of light of frequency ν has energy $h\nu$, where h is called Planck's constant. The packet is called the photon and the above rule is called Planck's law (see Box 3.4).

photographic magnitude See text (p. 19).

plasma Gas containing ionized atoms or molecules and free electrons.

Planck's law See *photon* and Box 3.4.

positron Dirac's investigations established the existence of 'anti-matter' which has the property of annihilating matter and giving rise to radiation. The positron is the opposite of the electron in this respect. It is positively charged and has the same mass as the electron, but is made of antimatter.

principle of least action A way of deducing new laws of physics. The 'action' is a mathematical entity defined in terms of physical quantities. The principle states that, in Nature, these quantities are related in such a way that, if they are changed slightly, the resulting change in the action is zero. This property is usually sufficient to deduce the relationship between the physcial quantities.

proper motion Motion of heavenly bodies perpendicular to the line of sight. (See Box 3.1.)

proton Electrically charged constituent of the atomic nucleus. The proton and the electron have equal and opposite charges but the proton has a mass about 1836·12 times the mass of the electron.

quantum physics Physics based on quantum theory. This theory imposes certain fundamental limitations on measurements of physical quantities and leads in many cases to a discrete behaviour of matter where classical (pre-quantum) physics predicted a continuous behaviour.

QSO (quasi-stellar object), *Quasar* Compact extragalactic object which presents a star-like appearance in spite of being much more (million times or more) massive than a star. A quasar is a QSO which also emits radio waves.

radian Measure of an angle. π radians make 180°, i.e. one radian equals approximately $57\frac{1}{4}°$.

red giant A star of enormous size compared to the Sun but having reddish colour. If a red giant star is located at the Sun, its outer surface may swallow up the inner planets, as well as the Earth, and even Mars.

red-shift See text for a detailed discussion (p. 74).

relativity: special and general Theories invented by Albert Einstein. The special theory (1905) revolutionized the concepts of space and time. It showed how they are related to the motion of the observer. The theory predicted several startling results most of which have been experimentally verified. The general theory, which followed 10 years later, attempted to explain the phenomenon of gravitation by another radical concept, that of curvature of space and time (see details on p. 152 of text).

sink and source Concepts in thermodynamics. The source is where energy originates, and the sink is where it is deposited. The subject of thermodynamics deals with the modes of transfer of energy from source to sink, and with the properties of the source and the sink themselves.

space–time The three dimensions of space and one of time became closely linked when Einstein formulated the theory of relativity (see *relativity*). 'Space–time' is used to represent the combination of space and time.

space–time diagrams Diagram showing the behaviour of a system in space and time.

statistical mechanics This is an attempt to understand the irreversible behaviour of macroscopic systems in terms of the statistical description of the large number of microscopic constituents making up the systems. In the statistical description details of the individual members are lost and only averages of ensembles are left.

theoretical physics Part of physics which seeks to explain the outcome of scientific experiments or the observations of natural phenomena in terms of mathematical models based on certain fundamental laws of Nature.

thermodynamics Science dealing with exchange of heat energy and its relation to work and the mechanical behaviour of physical systems.

thermodynamic equilibrium The ultimate state of disorder of a physical system. In the equilibrium state the emission and absorption of heat reach a balance and the 'entropy' of the system attains its maximum value.

thermonuclear reactions Reactions which take place when different atomic nuclei are brought together at high temperatures. The reactions result in changes of structure (break-up or fusion) of participating nuclei accompanied by a release or absorption of heat.

thought experiment Not an actual experiment but an imaginary one in which the experimental set up is assumed without worrying about how this can be done in real life. The outcome of the experiment can then be followed by reasoning and by using physical laws.

visual magnitude See text (p. 19).

waves Periodic change in a physical quantity with respect to spatial displacement and passage of time is called a wave. In a plane wave the change in the physical quantity takes place only in one direction of space. In a spherical wave the disturbance has a centre at a point in space; at all points in space at the same distance from the central point the disturbance is the same at any given moment of time.

wavelength The distance between two successive peaks or troughs of a wave.

Weyl postulate See text (p. 119).

white dwarf Star of highly dense matter in the form of electrons and protons. The density of matter in a white dwarf may be as high as -10^8 kg m^{-3} (see p. 39).

X-ray Electromagnetic radiation in the wavelength range of 0·03 Å to ~ 3 Å. This corresponds to an energy range of ~ 4–4000 keV. (See Box 3.4.)

A List of Books for Further Reading

The following is a list of books which deal with some of the topics discussed in this book. The list, by no means a comprehensive one, is arranged in a spectrum of increasing technicality. The present book lies somewhere near the centre of this spectrum.

The Individual and the Universe by A. C. B. Lovell. Based on the B.B.C. Reith Lectures (Harper, London, 1959).

The Nature of the Universe by C. W. Kilmister. A description of the development of astronomy from primitive days to the modern times (Thames and Hudson, London, 1971).

The Character of Physical Law by R. P. Feynman. Based on lectures given at Cornell University, U.S.A., this book deals with the requirements and the outcome of physical laws (B.B.C., London, 1965).

Galaxies, Nuclei and Quasars by F. Hoyle. A discussion of some problems of extragalactic astronomy (Heinemann, London, 1965).

Cosmology Today ed. Laurie John. A collection of articles by several authors dealing with the structure of the Universe, origin of elements, black holes, etc. (B.B.C., London, 1973).

Astronomy and Cosmology by F. Hoyle. A college-level textbook dealing with many important aspects of the subject (W. H. Freeman, San Fransisco, 1975).

Survey of the Universe by D. H. Menzel, F. L. Whipple, and G. De-Vaucouleurs. A college-level textbook dealing with astronomical topics (Prentice-Hall, New York, 1970).

Cosmology by H. Bondi. A college-level book dealing with the theoretical and observational aspects of cosmology (Cambridge University Press, 1960).

The Nature of Time ed. T. Gold. Based on a discussion between physicists and philosophers of science on the intriguing aspects of time (Cornell University Press, 1967).

Action at a Distance in Physics and Cosmology by F. Hoyle and J. V. Narlikar. A technical description of action-at-a-distance theories of electromagnetism and gravitation and their relation to the arrow of time (W. H. Freeman, San Francisco, 1974).

Index